The

Big

Switch

Also by Nicholas Carr

Does IT Matter?

The

Big

Switch

Rewiring the World, From Edison to Google

Nicholas Carr

W. W. NORTON & COMPANY

New York   London

Copyright © 2008 by Nicholas Carr

For information about permission to reproduce
selections from this book, write to Permissions,
W. W. Norton & Company, Inc., 500 Fifth Avenue, New York, NY 10110

For information about special discounts for bulk purchases, please contact
W. W. Norton Special Sales at specialsales@wwnorton.com or 800-233-4830

Manufacturing by Courier Westford
Book design by Chris Welch
Production manager: Anna Oler

Library of Congress Cataloging-in-Publication Data

Carr, Nicholas G., 1959–
The big switch : rewiring the world, from Edison to Google / Nicholas
Carr. — 1st ed.
p. cm.
Includes bibliographical references and index.
ISBN 978-0-393-06228-1 (hardcover)
1. Computers and civilization. 2. Information technology—
Social aspects. 3. Technological innovations. 4. Internet. I. Title.
QA76.9.C66C38 2008
303.48'34—dc22
2007038084

W. W. Norton & Company, Inc.
500 Fifth Avenue, New York, N.Y. 10110
www.wwnorton.com

W. W. Norton & Company Ltd.
Castle House, 75/76 Wells Street, London W1T 3QT

1 2 3 4 5 6 7 8 9 0

One lingered long among the dynamos,
for they were new, and they gave to history a new phase.
—*Henry Adams*

Contents

A Doorway in Boston

I T WAS A blustery November day, cold but bright, and I was lost. The directions I'd printed off the Internet weren't doing me any good. The road map that had looked so simple on my computer screen had turned into a ball of real-world confusion—thanks to Boston's cow-path roads and its plague of twisted street signs. As the digits of my dashboard clock clicked past the scheduled time for my lunch meeting, I decided I'd have better luck on foot. I pulled into an open parking space across from the high green walls of Fenway Park, got out of the car, and asked a passerby for directions. He pointed me to a nearby street and, at last able to follow the twists and turns of my MapQuest printout, I soon arrived at the right place: a hulking gray building at the end of a litter-strewn side street.

At least I thought it was the right place. I was looking for a company named VeriCenter, but there was no name on the building— just a beat-up little sign with a street number hanging from a post above a heavy steel door. I double-checked the address: it was definitely the right number. So I pushed the door open and walked into the world's most unwelcoming entryway: no furniture, no window, no company directory, no nothing. Just a black phone without a keypad on the wall beside another heavy steel door.

I lifted the phone and a man's voice came on the line. I gave him my name and the name of the person I'd come to meet, and he buzzed me through—into a second entryway, nearly as barren as the first. The man, a security guard, sat behind a metal desk. He put my driver's license through a tiny scanner, printed a blurry image of my face onto a visitor's pass, then had me sit in a chair beside an elevator. Someone would be down in a minute, he said. By this time, I was starting to wish I had stuck to my guns and turned down the meeting. A guy from VeriCenter's PR firm had been sending me emails for some time, and I'd been diligently forwarding them into the electronic trash bin. But when he managed to get hold of me by phone, I gave in and agreed to a meeting. Here I was, then perched on an uncomfortable chair in what appeared to be a dilapidated factory on the Friday before Thanksgiving in 2004.

To be honest, I found it pretty odd that the VeriCenter folks had been so intent on meeting me in the first place. I didn't know much about the company—it had been founded toward the end of the great dotcom boom, the publicist had told me, and it was headquartered in Houston—but I did know that it was in the information technology business, and most people in the IT field kept their distance from me. I was the "IT Doesn't Matter" guy. That was the title of an article I'd written a year and a half earlier—in May 2003—for the *Harvard Business Review*. I had argued that despite the many grand claims made about the power of corporate computer systems, they weren't actually all that important to a company's success. They were necessary—you couldn't operate without them—but most systems had become so commonplace that they no longer provided one company with an edge over its competitors. Whenever somebody did something new with a computer, everyone else soon followed suit. Strategically speaking, information technology had become inert. It was just another cost of doing business.

One reporter called the article "the rhetorical equivalent of a 50-megaton smart bomb." For months afterward, the high and the mighty of the technology world went out of their way to attack my heretical idea. Steve Ballmer, Microsoft's chief executive, declared it "hogwash." Carly Fiorina, then the head of Hewlett–Packard, said I was "dead wrong." Speaking at a big tech conference, Intel's CEO, Craig Barrett, boomed at the audience, "IT matters a whole lot!" The controversy even found its way into the popular press. *Newsweek* dubbed me "the technology world's Public Enemy No. 1." When the Harvard Business School Press published an expanded version of the article as a book, the industry went through a whole new round of hysterics.

So, as you might imagine, I wasn't exactly used to being invited to lunch by computer companies.

THE ELEVATOR OPENED and out stepped Jennifer Lozier, Veri-Center's smartly dressed marketing director. She escorted me up to a conference room and introduced me to a handful of her colleagues, including one of VeriCenter's founders, Mike Sullivan. A born entrepreneur, Sullivan could hardly contain his enthusiasm. He was holding a copy of my book, a couple of dozen Post-It notes sprouting from its pages. "When I read this," he said, "I knew I had to meet you. We're doing exactly what you write about." He tapped the cover of the book. "This is our business."

I was puzzled. Why would an IT company embrace the idea that IT doesn't matter?

Sullivan explained that he used to work as a general manager for Microsoft but that in 1999 he left to help launch VeriCenter because he wanted to pioneer a whole new way of supplying information technology to companies. He was convinced that, instead of buying and running their own computers and software, businesses in the future would just plug into the Internet and get all the data process-ing they needed, served up by outside utilities for a simple monthly fee. In my book, I'd compared information technology to electricity. VeriCenter, said Sullivan, was taking the next logical step: to actu-ally supply it like electricity, through a socket in the wall.

After a quick lunch and the obligatory PowerPoint presentation about the company, Sullivan said he wanted to give me a tour of "the data center." He led me back downstairs, through a hallway, to yet another door—this one constructed of steel mesh. The secu-rity guard diligently checked our badges before unlocking the door with a keycard chained to his belt. He escorted us inside.

Passing through that door was like entering a new world. The building may have looked like an old factory from the outside, but hidden inside was something altogether different—something not from the industrial past but the digital future. Stretching out before me, illuminated by the steady, sterile light of a thousand fluorescent bulbs, was a room the size of a city block, and it was filled with big computers. They stood in long rows inside locked cages, bearing the familiar logos of companies like IBM, Sun Microsystems, Dell, and HP. There didn't seem to be any other people in the room, just the machines, their fans humming and their red and green LEDs pulsing placidly as billions of bits of data streamed through their microprocessors. Overhead, big vents sucked up the heat from all the chips while other vents pumped in cool, filtered air.

Sullivan led me through the computers to a pair of side rooms that each held a huge Caterpillar diesel generator capable of pump-ing out two megawatts of electricity. With the fuel stored on site, he explained, the generators could keep the center operating for more than three days in the event of a power failure. He showed me another room that was packed from floor to ceiling with industrial-sized batteries, a second backup for briefer outages. We then walked to a corner of the facility where a fat pipe came through the wall. Holding a bundle of fiber-optic cables, it was the Internet connection that linked this roomful of computers to the dozens of businesses that used the data center to run their software and store their data. These companies no longer had to house and maintain their own gear or install and troubleshoot their own software. They just had to connect their offices, via the Internet, to the machines in this room. VeriCenter took care of the rest.

As I stood there surveying the data center, I might as well have been a cartoon figure with a big lightbulb flashing above my head. I realized that what I was standing in was a prototype of a new kind of power plant—a computing plant that would come to power our information age the way great electric plants powered the industrial age. Connected to the Net, this modern dynamo would deliver into our businesses and homes vast quantities of digitized information and data-processing might. It would run all the complicated software programs that we used to have to install on our own little computers. And, just like the earlier electric dynamos, it would operate with an efficiency never possible before. It would turn computing into a cheap, universal commodity.

"This really is a utility," I said to Sullivan.

He nodded, grinning. "This is the future."

One

Machine

... and likewise all parts of the system
must be constructed with reference to all other parts,
since, in one sense, all the parts form one machine.

—*Thomas Edison*

CHAPTER 1

Burden's Wheel

I N 1851, IN a field beside an ironworks in upstate New York, Henry Burden built a magnificent machine. Resembling a giant bicycle wheel, with dozens of thick iron spokes radiating from a massive central hub, Burden's machine was the largest industrial water-wheel in the country and the most powerful anywhere. Fed by a fast-moving spout of water diverted from the nearby Wynantskill River, the 60-foot-tall, 250-ton behemoth could produce as much as 500 horsepower when turning at its top speed of two and a half times a minute. The power was harnessed and delivered to the drill presses, grinding wheels, forge hammers, and lathes used by Burden's workers through an intricate system of gears, belts, and pulleys.

Henry Burden had a genius for industrial invention. A Scottish engineer, he emigrated to the United States in 1819, at the age of twenty-eight, to take a position with a farming-tool manufacturer in Albany. Within months he had invented the country's first culti-vator for preparing seed beds and designed an improved version of a plow. Three years later, he moved to the nearby town of Troy to manage the Troy Iron and Nail Factory, which he would eventually buy and rename the Burden Iron Works. It didn't take him long to see the boon in the plant's location near the junction of the Hudson River and the freshly dug Erie Canal. If he could boost the factory's output, he would be able to ship its products to new markets throughout the Northeast and Midwest. He set to work mechanizing what had for centuries been a local industry built on the handiwork of blacksmiths and other craftsmen. Within a dozen years, he had created machines that automated the production of nails and railroad spikes, and in 1835 he invented the Burden Horseshoe Machine, an ingenious contraption that turned bars of iron into finished shoes at the rate of one a second. In his spare time, Burden also managed to design a large ocean-going steamboat that became the model for many subsequent ferries and cruise ships.

But Burden's greatest creation, the one that made him rich as well as famous, was his wheel. Dubbed "the Niagara of Water-wheels" by a local poet, its unprecedented size and power provided the Burden Iron Works with a decisive advantage over other manufacturers. The company was able to expand the yield and efficiency of its factory, producing more shoes, spikes, and other goods with fewer workers and in less time than rivals could manage. It won a contract to supply nearly all the horseshoes used by the Union army during the Civil War, and it became one of the major suppliers of spikes to American railroads as they extended their lines across the country.

For Burden, the efficient production of mechanical power turned out to be every bit as important to his company's success as the skill of his workers and even the quality of his products. Like other factory owners of the time, he was as much in the business of manufacturing energy as manufacturing goods.

But a visitor to the Burden Iron Works in the early years of the twentieth century would have come upon a surprising sight. The great waterwheel stood idle in the field, overgrown with weeds and quietly rusting away. After turning nonstop for fifty years, it had been abandoned. Manufacturers didn't have to be in the power-generation business anymore. They could run their machines with electric current generated in distant power plants by big utilities and fed to their factories over a network of wires. With remarkable speed, the new utilities took over the supply of industrial power. Burden's wheel and thousands of other private waterwheels, steam engines, and electric generators were rendered obsolete.

What made large-scale electric utilities possible was a series of scientific and engineering breakthroughs—in electricity generation and transmission as well as in the design of electric motors—but what ensured their triumph was not technology but economics. By supplying electricity to many buyers from central generating stations, the utilities achieved economies of scale in power production that no individual factory could match. It became a competitive necessity for manufacturers to hook their plants up to the new electric grid in order to tap into the cheaper source of power. The success of the utilities fed on itself. As soon as they began to supply current to factories, they were able to expand their generating capacity and scale economies even further, achieving another great leap in efficiency. The price of electricity fell so quickly that it soon became possible for nearly every business and household in the country to afford electric power.

The commercial and social ramifications of the democratization of electricity would be hard to overstate. Electric light altered the rhythms of life, electric assembly lines redefined industry and work, and electric appliances brought the Industrial Revolution into the home. Cheap and plentiful electricity shaped the world we live in today. It's a world that didn't exist a mere hundred years ago, and yet the transformation that has played out over just a few generations has been so great, so complete, that it has become almost impossible for us to imagine what life was like before electricity began to flow through the sockets in our walls.

TODAY, WE'RE IN the midst of another epochal transformation, and it's following a similar course. What happened to the generation of power a century ago is now happening to the processing of information. Private computer systems, built and operated by individual companies, are being supplanted by services provided over a common grid—the Internet—by centralized data-processing plants. Computing is turning into a utility, and once again the economic equations that determine the way we work and live are being rewritten.

For the past half century, since the first mainframe computer was installed in a corporate data center, businesses have invested many trillions of dollars in information technology. They've assembled hardware and software into ever more complex systems to automate nearly every facet of their operations, from the way they buy materials and supplies to the way they manage their employees to the way they deliver their products to customers. They've housed those systems on-site, in their own plants and offices, and they've maintained them with their own staffs of technicians. Just as Henry Burden and other manufacturers competed in part on the sophistication of their power systems, so modern companies have competed on the sophistication of their computer systems. Whatever their main business might be, they've had no choice but to also be in the business of data processing.

Until now.

Capitalizing on advances in the power of microprocessors and the capacity of data storage systems, fledgling utilities are beginning to build massive and massively efficient information-processing plants, and they're using the broadband Internet, with its millions of miles of fiber-optic cable, as the global grid for delivering their services to customers. Like the electric utilities before them, the new computing utilities are achieving economies of scale far beyond what most companies can achieve with their own systems.

Seeing the economic advantages of the utility model, corporations are rethinking the way they buy and use information technology. Rather than devoting a lot of cash to purchasing computers and software programs, they're beginning to plug into the new grid. That shift promises not only to change the nature of corporate IT departments but to shake up the entire computer industry. Big tech companies—Microsoft, Dell, Oracle, IBM, and all the rest—have made tons of money selling the same systems to thousands of companies. As computing becomes more centralized, many of those sales will dry up. Considering that businesses spend well over a trillion dollars a year on hardware and software, the ripple effects will be felt throughout the world economy.

But this is hardly just a business phenomenon. Many of the most advanced examples of utility computing are aimed not at companies but at people like you and me. The best example of all is probably Google's search engine. Think about it: what is Google but a giant information utility? When you need to search the Internet, you use your Web browser to connect to the vast data centers that Google has constructed in secret locations around the world. You type in a keyword, and Google's network of hundreds of thousands of interlinked computers sorts through a database of billions of Web pages, draws out the few thousand that best match up with your keyword, arranges them in order of relevance, and shoots the results back through the Internet to your screen—usually in less than a second. That amazing computing feat, which Google repeats hundreds of millions of times a day, doesn't happen inside your PC. It *couldn't* happen inside your PC. Rather, it happens miles away, maybe on the other side of the country, maybe even on the other side of the globe. Where's the computer chip that processed your last Google search? You don't know, and you don't care—any more than you know or care which generating station produced the kilowatts that light the lamp on your desk.

All historical models and analogies have their limits, of course, and information technology differs from electricity in many important ways. But beneath the technical differences, electricity and computing share deep similarities—similarities that are easy for us to overlook today. We see electricity as a "simple" utility, a standardized and unremarkable current that comes safely and predictably through outlets in our walls. The innumerable applications of electric power, from televisions and washing machines to machine tools and assembly lines, have become so commonplace that we no longer consider them to be elements of the underlying technology—they've taken on separate, familiar lives of their own.

It wasn't always so. When electrification began, it was an untamed and unpredictable force that changed everything it touched. Its applications were as much a part of the technology as the dynamos, the power lines, and the current itself. As with today's computer systems, all companies had to figure out how to apply electricity to their own businesses, often making sweeping changes to their organizations and processes. As the technology advanced, they had to struggle with old and often incompatible equipment—the "legacy systems," to use a modern computer term, that can lock businesses into the past and impede progress—and they had to adapt to customers' changing needs and expectations. Electrification, just like computerization, led to complex, far-reaching, and often bewildering changes for individual companies and entire industries—and, as households began to connect to the grid, for all of society.

At a purely economic level, the similarities between electricity and information technology are even more striking. Both are what economists call general purpose technologies. Used by all sorts of people to do all sorts of things, they perform many functions rather than just one or a few. General purpose technologies, or GPTs, are best thought of not as discrete tools but as platforms on which many different tools, or applications, can be constructed. Compare the electric system to the rail system. Once railroad tracks are laid, you can pretty much do only one thing with them: run trains back and forth carrying cargo or passengers. But once you set up an electric grid, it can be used to power everything from robots in factories to toasters on kitchen counters to lights in classrooms. Because they're applied so broadly, GPTs offer the potential for huge economies of scale—if their supply can be consolidated.

That's not always possible. Steam engines and waterwheels were general purpose technologies that didn't lend themselves to centralization. They had to be located close to the point where their power was used. That's why Henry Burden had to build his wheel right next to his factory. If he had built it even a few hundred yards away, all the energy produced by the wheel would have been consumed in turning the long shafts and belts required to convey the energy to the factory. There wouldn't have been any left to power workers' machines.

But electricity and computing share a special trait that makes them unique even among the relatively small set of general purpose technologies: they can both be delivered efficiently from a great distance over a network. Because they don't have to be produced locally, they can achieve the scale economies of central supply. Those economies, though, can take a long time to be fully appreciated and even longer to be fully exploited. In the early stages of a GPT's development, when there are few technical standards and no broad distribution network, the technology is impossible to furnish centrally. Its supply is by necessity fragmented. If a company wants to tap into the power of the technology, it has to purchase the various components required to supply it, install those components at its own site, cobble them together into a working system, and hire a staff of specialists to keep the system running. In the early days of electrification, factories had to build their own generators if they wanted to use the power of electricity—just as today's companies have had to set up their own information systems to use the power of computing.

Such fragmentation is wasteful. It imposes large capital investments and heavy fixed costs on firms, and it leads to redundant expenditures and high levels of overcapacity, both in the technology itself and in the labor force operating it. The situation is ideal for the suppliers of the components of the technology—they reap the benefits of overinvestment—but it's not sustainable. Once it becomes possible to provide the technology centrally, large-scale utility suppliers arise to displace the private providers. It may take decades for companies to abandon their proprietary supply operations and all the investments they represent. But in the end the savings offered by utilities become too compelling to resist, even for the largest enterprises. The grid wins.

AT A CONFERENCE in Paris during the summer of 2004, Apple introduced an updated version of its popular iMac computer. Since its debut in 1998, the iMac had always been distinguished by its unusual design, but the new model was particularly striking. It appeared to be nothing more than a flat-panel television, a rectangular screen encased in a thin block of white plastic and mounted on an aluminum pedestal. All the components of the computer itself—the chips, the drives, the cables, the connectors—were hidden behind the screen. The advertising tagline wittily anticipated the response of prospective buyers: "Where did the computer go?"

But the question was more than just a cute promotional pitch. It was, as well, a subtle acknowledgment that our longstanding idea of a computer is obsolete. While most of us continue to depend on personal computers both at home and at work, we're using them in a very different way than we used to. Instead of relying on data and software that reside inside our computers, inscribed on our private hard drives, we increasingly tap into data and software that stream through the public Internet. Our PCs are turning into terminals that draw most of their power and usefulness not from what's inside them but from the network they're hooked up to—and, in particular, from the other computers that are hooked up to that network.

The change in the way we use computers didn't happen overnight. Primitive forms of centralized computing have been around for a long time. In the mid-1980s, many early PC owners bought modems to connect their computers over phone lines to central databases like CompuServe, Prodigy, and the Well—commonly known as "bulletin boards"—where they exchanged messages with other subscribers. America Online popularized this kind of online community, greatly expanding its appeal by adding colorful graphics as well as chat rooms, games, weather reports, magazine and newspaper articles, and many other services. Other, more specialized databases were also available to scholars, engineers, librarians, military planners, and business analysts. When, in 1990, Tim Berners-Lee invented the World Wide Web, he set the stage for the replacement of all those private online data stores with one vast public one. The Web popularized the Internet, turning it into a global bazaar for sharing digital information. And once easy-to-use browsers like Netscape Navigator and Internet Explorer became freely available in the mid-1990s, we all went online in droves.

Through the first decade of its existence, however, the World Wide Web was a fairly prosaic place for most of us. We used it mainly as a giant catalog, a collection of "pages" bound together with hyperlinks. We "read" the Web, browsing through its contents in a way that wasn't so different from the way we'd thumb through a pile of magazines. When we wanted to do real work, or play real games, we'd close our Web browser and launch one of the many programs installed on our own hard drive: Microsoft Word, maybe, or Aldus Pagemaker, or Encarta, or Myst.

But beneath the Web's familiar, page-like surface lay a set of powerful technologies, including sophisticated protocols for describing and transferring data, that promised not only to greatly magnify the usefulness of the Internet but to transform computing itself. These technologies would allow all the computers hooked up to the Net to act, in effect, as a single information-processing machine, easily sharing bits of data and strings of software code. Once the technologies were fully harnessed, you'd be able to use the Internet not just to look at pages on individual sites but to run sophisticated software programs that might draw information from many sites and databases simultaneously. You'd be able not only to "read" from the Internet but to "write" to it as well—just as you've always been able to read from and write to your PC's hard drive. The World Wide Web would turn into the World Wide Computer.

This other dimension of the Internet was visible from the start, but only dimly so. When you ran a Web search on an early search engine like AltaVista, you were running a software program through your browser. The code for the software resided mainly on the computer that hosted AltaVista's site. When you did online banking, shifting money between a checking and a savings account, you were also using a utility service, one that was running on your bank's computer rather than your own. When you used your browser to check your Yahoo or Hotmail email account, or track a FedEx shipment, you were using a complicated application running on a distant server computer. Even when you used Amazon.com's shopping-cart system to order a book—or when you subsequently posted a review of that book on the Amazon site—you were tapping into the Internet's latent potential.

For the most part, the early utility services were rudimentary, involving the exchange of a small amount of data. The reason was simple: more complex services, the kind that might replace the software on your hard drive, required the rapid transfer of very large quantities of data, and that just wasn't practical with traditional, low-speed dial-up connections. Running such services would quickly overload the capacity of telephone lines or overwhelm your modem. Your PC would grind to a halt. Before sophisticated services could proliferate, a critical mass of people had to have high-speed broadband connections. That only began to happen late in the 1990s during the great dotcom investment boom, when phone and cable companies rushed to replace their copper wires with optical fibers—hair-thin strands of glass that carry information as pulses of light rather than electric currents—and retool their networks to handle virtually unlimited quantities of data.

The first clear harbinger of the second coming of the Internet—what would eventually be dubbed Web 2.0—appeared out of nowhere in the summer of 1999. It came in the form of a small, free software program called Napster. Written over a few months by an eighteen-year-old college dropout named Shawn Fanning, Napster allowed people to share music over the Internet in a whole new way. It scanned the hard drive of anyone who installed the program, and then it created, on a central server computer operated by Fanning, a directory of information on all the song files it found, cataloging their titles, the bands that performed them, the albums they came from, and their audio quality. Napster users searched this directory to find songs they wanted, which they then downloaded directly from other users' computers. It was easy and, if you had a broadband connection, it was fast. In a matter of hours, you could download hundreds of tunes. It's no exaggeration to say that, at Napster's peak, almost every work of popular music that had ever been digitally encoded onto a compact disk—and many that had never appeared on a disk—could be found and downloaded for free through the service.

Napster, not surprisingly, became wildly popular, particularly on college campuses where high-speed Net connections were common. By early 2001, according to an estimate by market researcher Media Metrix, more than 26 million people were using the service, and they were spending more than 100 million hours a month exchanging music files. Shawn Fanning's invention showed the world, for the first time, how the Internet could allow many computers to act as a single shared computer, with thousands or even millions of people having access to the combined contents of previously private databases. Although every user had to install a little software program on his own PC, the real power of Napster lay in the network itself—in the way it created a central file-management system and the way it allowed data to be transferred easily between computers, even ones running on opposite sides of the planet.

There was just one problem. It wasn't legal. The vast majority of the songs downloaded through Napster were owned by the artists and record companies that had produced them. Sharing them without permission or payment was against the law. The arrival of Napster had turned millions of otherwise law-abiding citizens into digital shoplifters, setting off the greatest, or at least the broadest, orgy of looting in history. The musicians and record companies fought back, filing lawsuits charging Fanning's company with copyright infringement. Their legal counterattack culminated in the closing of the service in the summer of 2001, just two years after it had launched.

Napster died, but the business of supplying computing services over the Internet exploded in its wake. Many of us now spend more time using the new Web services than we do running traditional software applications from our hard drives. We rely on the new utility grid to connect with our friends at social networks like MySpace and Facebook, to manage our photo collections at sites like Flickr and Photobucket, to create imaginary selves in virtual worlds like World of Warcraft and Disney's Club Penguin, to watch videos through services like YouTube and Joost, to write blogs with Word-Press or memos with Google Docs, to follow breaking news through feed readers like Rojo and Bloglines, and to store our files on "virtual hard drives" like Omnidrive and Box.

All these services hint at the revolutionary potential of the information utility. In the years ahead, more and more of the information-processing tasks that we rely on, at home and at work, will be handled by big data centers located out on the Internet. The nature and economics of computing will change as dramatically as the nature and economics of mechanical power changed in the early years of the last century. The consequences for society—for the way we live, work, learn, communicate, entertain ourselves, and even think—promise to be equally profound. If the electric dynamo was the machine that fashioned twentieth-century society—that made us who we are—the information dynamo is the machine that will fashion the new society of the twenty-first century.

LEWIS MUMFORD, IN his 1970 book *The Pentagon of Power*, the second volume of his great critique of technology *The Myth of the Machine*, made an eloquent case against the idea that technological progress determines the course of history. "Western society," he wrote, "has accepted as unquestionable a technological imperative that is quite as arbitrary as the most primitive taboo: not merely the duty to foster invention and constantly to create technological novelties, but equally the duty to surrender to these novelties unconditionally, just because they are offered, without respect to their human consequences." Rather than allowing technology to control us, Mumford implied, we can control technology—if only we can muster the courage to exert the full power of our free will over the machines we make.

It's a seductive sentiment, one that most of us would like to share, but it's mistaken. Mumford's error lay not in asserting that as a society we pursue and embrace technological advances with little reservation. That's hard to dispute. His error lay in suggesting that we might do otherwise. The technological imperative that has shaped the Western world is not arbitrary, nor is our surrender to it discretionary. The fostering of invention and the embrace of the new technologies that result are not "duties" that we have somehow chosen to accept. They're the consequences of economic forces that lie largely beyond our control. By looking at technology in isolation, Mumford failed to see that the path of technological progress and its human consequences are determined not simply by advances in science and engineering but also, and more decisively, by the influence of technology on the costs of producing and consuming goods and services. A competitive marketplace guarantees that more efficient modes of production and consumption will win out over less efficient ones. That's why Henry Burden built his wheel, and it's why that wheel was left to rust a few decades later. Technology shapes economics, and economics shapes society. It's a messy process—when you combine technology, economics, and human nature, you get a lot of variables—but it has an inexorable logic, even if we can trace it only in retrospect. As individuals, we may question the technological imperative and even withstand it, but such acts will always be lonely and in the end futile. In a society governed by economic trade-offs, the technological imperative is precisely that: an imperative. Personal choice has little to do with it.

We see the interplay of technology and economics most clearly at those rare moments when a change takes place in the way a resource vital to society is supplied, when an essential product or service that had been supplied locally begins to be supplied centrally, or vice versa. Civilization itself emerged when food production, decentralized in primitive hunter-gatherer societies, began to be centralized with the introduction of the technologies of agriculture. Changes in the supply of other important resources—resources as diverse as water, transportation, the written word, and government—also altered the economic trade-offs that shape society. A hundred years ago, we arrived at such a moment with the technologies that extend man's physical powers. We are at another such moment today with the technologies that extend our intellectual powers.

The transformation in the supply of computing promises to have especially sweeping consequences. Software programs already control or mediate not only industry and commerce but entertainment, journalism, education, even politics and national defense. The shock waves produced by a shift in computing technology will thus be intense and far-reaching. We can already see the early effects all around us—in the shift of control over media from institutions to individuals, in people's growing sense of affiliation with "virtual communities" rather than physical ones, in debates over the security of personal information and the value of privacy, in the export of the jobs of knowledge workers, even in the growing concentration of wealth in a small slice of the population. All these trends either spring from or are propelled by the rise of Internet-based computing. As information utilities grow in size and sophistication, the changes to business and society—and to ourselves—will only broaden. And their pace will only accelerate.

Many of the characteristics that define American society came into being only in the aftermath of electrification. The rise of the middle class, the expansion of public education, the flowering of mass culture, the movement of the population to the suburbs, the shift from an industrial to a service economy—none of these would have happened without the cheap current generated by utilities. Today, we think of these developments as permanent features of our society. But that's an illusion. They're the by-products of a particular set of economic trade-offs that reflected, in large measure, the technologies of the time. We may soon come to discover that what we assume to be the enduring foundations of our society are in fact only temporary structures, as easily abandoned as Henry Burden's wheel.

CHAPTER 2

The Inventor and His Clerk

THOMAS EDISON was tired. It was the summer of 1878, and he'd just spent a grueling year perfecting and then promoting his most dazzling invention yet, the tinfoil phonograph. He needed a break from the round-the-clock bustle of his Menlo Park laboratory, a chance to clear his mind before embarking on some great new technological adventure. When a group of his friends invited him to join them on a leisurely camping and hunting tour of the American West, he quickly agreed. The trip began in Rawlins, Wyoming, where the party viewed an eclipse of the sun, and then continued westward through Utah and Nevada, into Yosemite Valley, and on to San Francisco.

While traveling through the Rockies, Edison visited a mining site by the side of the Platte River. Seeing a crew of workers struggling with manual drills, he turned to a companion and remarked, "Why cannot the power of yonder river be transmitted to these men by electricity?" It was an audacious thought—electricity had yet to be harnessed on anything but the smallest scale—but for Edison audacity was synonymous with inspiration. By the time he returned east in the fall, he was consumed with the idea of supplying electricity over a network from a central generating station. His interest no longer lay in powering the drills of work crews in the wilderness, however. He wanted to illuminate entire cities. He rushed to set up the Edison Electric Light Company to fund the project and, on October 20, he announced to the press that he would soon be providing electricity to the homes and offices of New York City. Having made the grand promise, all he and his Menlo Park team had to do was figure out how to fulfill it.

Unlike lesser inventors, Edison didn't just create individual products; he created entire systems. He first imagined the whole, then he built the necessary pieces, making sure they all fit together seamlessly. "It was not only necessary that the lamps should give light and the dynamos generate current," he would later write about his plan for supplying electricity as a utility, "but the lamps must be adapted to the current of the dynamos, and the dynamos must be constructed to give the character of current required by the lamps, and likewise all parts of the system must be constructed with reference to all other parts, since, in one sense, all the parts form one machine." Fortunately for Edison, he had a good model at hand. Urban gaslight systems, invented at the start of the century, had been set up in many cities to bring natural gas from a central gasworks into buildings to be used as fuel for lamps. Light, having been produced by simple candles and oil lamps for centuries, had already become a centralized utility. Edison's challenge was to replace the gaslight systems with electric ones.

Electricity had, in theory, many advantages over gas as a source of lighting. It was easier to control, and because it provided illumination without a flame it was cleaner and safer to use. Gaslight by comparison was dangerous and messy. It sucked the oxygen out of rooms, gave off toxic fumes, blackened walls and soiled curtains, heated the air, and had an unnerving tendency to cause large and deadly explosions. While gaslight was originally "celebrated as cleanliness and purity incarnate," Wolfgang Schivelbusch reports in *Disenchanted Night,* his history of lighting systems, its shortcomings became more apparent as it came to be more broadly used. People began to consider it "dirty and unhygienic"—a necessary evil. Edison himself dismissed gaslight as "barbarous and wasteful." He called it "a light for the dark ages."

Despite the growing discontent with gas lamps, technological constraints limited the use of electricity for lighting at the time Edison began his experiments. For one thing, the modern incandescent lightbulb had yet to be invented. The only viable electric light was the arc lamp, which worked by sending a naked current across a gap between two charged iron rods. Arc lamps burned with such intense brightness and heat that you couldn't put them inside rooms or most other enclosed spaces. They were restricted to large public areas. For another thing, there was no way to supply electricity from a central facility. Every arc lamp required its own battery. "Like the candle and the oil lamp," Schivelbusch explains, "arc lighting was governed by the pre-industrial principle of a self-sufficient supply." However bad gaslight might be, electric light was no alternative.

To build his "one machine," therefore, Edison had to pursue technological breakthroughs in every major component of the system. He had to pioneer a way to produce electricity efficiently in large quantities, a way to transmit the current safely to homes and offices, a way to measure each customer's use of the current, and, finally, a way to turn the current into controllable, reliable light suitable for normal living spaces. And he had to make sure that he could sell electric light at the same price as gaslight and still turn a profit.

It was a daunting challenge, but he and his Menlo Park associates managed to pull it off with remarkable speed. Within two years, they had developed all the critical components of the system. They had invented the renowned Edison lightbulb, sealing a thin copper filament inside a small glass vacuum to create, as one reporter poetically put it, "a little globe of sunshine, a veritable Aladdin's lamp." They had designed a powerful new dynamo that was four times bigger than its largest precursor. (They named their creation the Jumbo, after a popular circus elephant of the time.) They had perfected a parallel circuit that would allow many bulbs to operate independently, with separate controls, on a single wire. And they had created a meter that would keep track of how much electricity a customer used. In 1881, Edison traveled to Paris to display a small working model of his system at the International Exposition of Electricity, held in the Palais de l'Industrie on the Champs-Elysées. He also unveiled blueprints for the world's first central generating station, which he announced he would construct in two warehouses on Pearl Street in lower Manhattan.

The plans for the Pearl Street station were ambitious. Four large coal-fired boilers would create the steam pressure to power six 125-horsepower steam engines, which in turn would drive six of Edison's Jumbo dynamos. The electricity would be sent through a network of underground cables to buildings in a square-mile territory around the plant, each of which would be outfitted with a meter. Construction of the system began soon after the Paris Exposition, with Edison often working through the night to supervise the effort. A little more than a year later, the plant had been built and the miles of cables laid. At precisely three o'clock in the afternoon on September 4, 1882, Edison instructed his chief electrician, John Lieb, to throw a switch at the Pearl Street station, releasing the current from one of its generators. As the *New York Herald* reported the following day, "in a twinkling, the area bounded by Spruce, Wall, Nassau and Pearl Streets was in a glow." The electric utility had arrived.

But running a utility was not what really interested Edison. The

Pearl Street station was, in his eyes, simply a proof of concept, an installation designed to demonstrate that his electric-light system would work. Edison's true business interest lay in franchising or licensing the patented system to other operators and then selling them the many components required to build and operate their plants. He organized a business empire to pursue his ambition. The Edison Company for Isolated Lighting licensed his system throughout the United States, while the Compagnie Continentale Edison and other affiliates did the same in Europe. The Edison Lamp Works produced lightbulbs. The Edison Machine Works manufactured dynamos. The Edison Electric Tube Company supplied the wiring. Yet another company sold various accessories. As demand for Edison's electric systems grew, so did his many-armed enterprise.

But the inventor's success also blinded him. Despite his visionary genius, he couldn't see beyond his licensing and components business. He had at first assumed that electric utilities would simply be a more attractive substitute for gas utilities: they would be relatively small, urban plants serving the lighting needs of nearby offices and homes. Indeed, since Edison's systems ran on direct current, which couldn't be transmitted far, they were unable to serve territories greater than a square mile. As the applications of electricity spread to factories and transit systems, Edison clung to his belief in small-scale, direct-current production. He assumed that industrial companies would build their own private generating plants with his plans and parts. Edison's pride in what he viewed as the perfection of his system reinforced this belief, but so did his economic interests. After all, the more small systems that were built, whether central stations or private plants, the more components he could sell. Edison had invented the first viable electric utility, but he couldn't envision the next logical step: the consolidation of electricity production into giant power plants and the creation of a national grid to share the power. The system that Edison had imagined, and then brought to life, came to be his imagination's cage.

It would take a very different man with a very different vision to fulfill the promise of the electric utility. It would take a man as talented in perfecting the economics of a technological system as Edison was in perfecting the technology itself. The irony is that the man's employer—and hero—was Edison himself.

ON THE EVENING of February 28, 1881, the ocean liner *City of Chester* pulled into the port of New York carrying a slight, near-sighted twenty-one-year-old English stenographer named Samuel Insull. He had been seasick for nearly the entire voyage, but that didn't dampen his spirits as he walked down the gangplank. He knew that he would soon realize his dream: to meet the legendary inventor Thomas Edison.

Insull was a serious and driven young man. Born into a family of temperance crusaders, he spent his boyhood poring over books with titles like *Lives of the Great Engineers* and *Self-Help*. From the start, he displayed, as his biographer Forrest McDonald describes, "a peculiar metabolic make-up. He invariably awoke early, abruptly, completely, bursting with energy; yet he gained momentum as the day wore on, and long into the night." Like Edison, Insull was a tireless, often monomaniacal worker—a human dynamo. He also shared Edison's gift for systems thinking, though it was business rather than mechanical systems that stirred Insull's passion. "Very early," writes McDonald, "he learned to see to the heart of relations between things, or between men and things and men and men, and to grasp the underlying principles so clearly that he could perceive ways to shift them around a bit and make them work the better." Though the abstractions of scholarship bored him, he had "a natural aptitude for quantitative, arithmetic analysis of what he saw—the accountant's way of viewing things."

When Insull was fourteen, he left school to take a job as an office boy at a London auction house. He learned shorthand from a colleague and soon took a second job, working nights as a stenographer for a newspaper editor. In his spare time, he taught himself bookkeeping, went to the opera, and read widely, filing everything away in his capacious memory. In 1878, just as he was turning nineteen, he happened upon a magazine article with a drawing of Thomas Edison. It was an event that would, as Insull recalled many years later, change his life:

> One evening I was going on the underground railway in London from my home to my work, where I was taking shorthand notes for a leading London editor, when I happened to pick up an old *Scribner's Monthly*. It contained a sketch of Mr. Edison in his laboratory at Menlo Park, the place where he carried on his early experiments in electric lighting. . . . I wrote an essay for the literary society of which I was a member on the subject of "The American Inventor, Thomas Alva Edison." Little did I think when I was looking up the information for that essay that my career would be made thousands of miles away under the name of the inventor who ultimately became one of the closest friends I have ever had.

Not long after writing his essay, Insull took a job as private secretary to a prominent banker named George Gouraud. It was a fortuitous move. Gouraud turned out to be the man overseeing Edison's European business affairs. Through his new boss, Insull met and struck up a friendship with Edison's chief engineer, Edward Johnson. Johnson became so impressed with Insull's intelligence and energy, not to mention his exhaustive knowledge of Edison's work, that he soon recommended that Edison bring the young man to America and hire him as his own secretary.

When Insull stepped off the *City of Chester*, Johnson was there, waiting to take him to the Manhattan offices of the Edison Electric Light Company. There, Insull was introduced to a harried and unshaven Edison, who immediately put his new assistant to work reviewing the company's complicated—and precarious—financing arrangements. Edison and Insull worked side by side through the night, and by dawn Insull had come up with a creative plan to borrow additional money using as collateral a package of Edison's European patents. "From that moment," reports McDonald, "Insull was Edison's financial factotum." He was more than that, actually. He was the clerk of the great inventor's works.

Insull played an instrumental role in keeping Edison's diverse and perpetually cash-starved operation running as demand for electric power grew. He supervised various parts of the Edison empire, reorganized its marketing and sales functions, traveled the country promoting the building of central stations, and negotiated deals with bankers and other financiers. In 1889, he oversaw the consolidation of Edison's manufacturing companies into the Edison General Electric Company and, three years later, played a central role in merging it with its largest competitor, Thomson–Houston, to become, simply, General Electric. But though Insull, at the age of just thirty-two, had become one of the highest-ranking executives of one of the world's most prestigious companies, he was unhappy with his position. He had studied every aspect of the power business, from the technology to the finances to the laws and regulations, and he yearned to be in charge. He had little interest in being a bureaucrat, no matter how elevated and well-paid, in a large and increasingly byzantine organization.

Even more important, his thinking about the electricity industry had diverged from that of his mentor. He had become convinced that running utilities would in the end be a more important business than manufacturing components for them. He had been following the rapid advances in electricity generation, transmission, and use, and he had begun to see beyond Edison's system to an entirely new model and role for central stations. In the spring of 1892, Insull was offered the presidency of the Chicago Edison Company, a small independent power producer serving just 5,000 customers. He accepted immediately. The move entailed a drastic cut in salary, from $36,000 to $12,000, but the compensation wasn't what mattered to him. He was looking to a farther horizon. At his farewell dinner in New York, he stood and, with fire in his eyes, promised that little Chicago Edison would grow to eclipse the great General Electric in size. The prediction, McDonald writes, "was so far-fetched as to be laughable—except that when Samuel Insull looked like that, nobody ever laughed."

What Insull had realized, or at least sensed, was that utility-supplied electricity could serve a far greater range of needs than it had up to then. Electricity could become a true general purpose technology, used by businesses and homeowners to run all kinds of machines and appliances. But for electricity and electric utilities to fulfill their destiny, the way power was produced, distributed, and consumed would need to be transformed. Just as Edison had to overcome many daunting challenges to weave together his utility system, Insull would have to do the same to reinvent that system. The biggest challenge of all would lie in convincing industrial businesses that they should stop producing their own power and instead buy it as a service from central plants. That would test all of Insull's skills as a businessman.

FROM THE TIME people first began using machines, they had no choice but to also produce the power necessary to run them. The original source of power was sheer brawn. As Louis C. Hunter writes in his *History of Industrial Power in the United States,* "During unrecorded millennia the muscles of men and animals had supplied the necessary motion for the earliest anonymous machines—the quern, the potter's wheel, the bow drill, the forge bellows, or the hand pump." Even as machines became more complex, it was still usually muscles that drove them. Horses tied to winches turned millstones to grind grains, saws to cut lumber, presses to bale cotton, and drills to dig tunnels and excavate quarries. "Uncounted men and animals," Hunter writes, "contributed most of the energy for the small enterprises that comprised the greater part of manufacturing industry before 1900."

But while muscle power was sufficient for small enterprises, it was not enough for larger ones. As the production of goods began to be centralized in factories, manufacturers required big, reliable, and controllable supplies of power to run their machinery. The first great source of industrial power was running water. Manufacturers would build their plants beside streams and rivers, harnessing the force of the flow with waterwheels and turning it into mechanical energy. Using water as a source of power had a long heritage, reaching back well before the Industrial Revolution. The Greeks and Romans had used waterwheels, and European farmers had been building rudimentary water-powered gristmills for centuries. When William the Conqueror surveyed England in 1066 for his Domesday book, he found thousands of such mills across the countryside.

During the 1800s, water-power systems grew much more sophisticated as they were adapted for use by larger factories. Hydraulic engineers worked to make waterwheels more efficient, introducing a series of design improvements. In addition to perfecting traditional wheels, like Henry Burden's goliath, they developed hydraulic turbines—powerful, fanlike wheels that also came to be widely adopted. There were also rapid advances in the design of dams, locks, and channels aimed at regulating water flow with the precision required to operate intricate and sensitive machinery.

Using water power had once been simple. A mill owner would contract with a local carpenter to make a basic wooden wheel with a driveshaft, and he'd put the wheel into a swiftly moving stream. Now, generating power was complicated and costly—and growing more so every day. Factory owners had to either learn hydraulics or hire experts who knew the science. They had to invest considerable capital into the construction and upkeep of their water-power systems, and they had to make difficult decisions about which kind of wheel to buy and what design to use for managing water flows. Once routine, choices about power generation could now make or break a company.

Complicating the matter further was the introduction of the second great technology for industrial power generation, the steam engine. Invented in the eighteenth century, steam engines transformed thermal energy into mechanical energy by boiling water to create steam that, as it expanded, pushed a piston or turned a turbine. Their great advantage was that they didn't require flowing water—they freed manufacturers from having to build factories beside streams and rivers. Their great disadvantage was that they were even more expensive to operate than waterwheels. They required a lot of fuel, in the form of coal or wood, to keep the water boiling.

As with hydraulic systems, steam technology progressed rapidly, with inventors and engineers around the world competing to create more efficient and reliable engines. The advances in the generation of power were matched by advances in its transmission. With industrial production, it was no longer enough to connect a waterwheel or a steam engine directly to a single machine like a millstone. Power had to be distributed to many different devices spread throughout a factory or even across several buildings on one site. This required the construction of "millwork"— assemblies of gears, shafts, belts, and pulleys for transmitting and regulating power.

As factories expanded and production processes became more involved, millwork in turn grew astonishingly elaborate. Factory owners had to hire architects to design the systems and skilled technicians to maintain them. A visitor to a British plant in the 1870s reported that the interior "present[ed] a bewildering appearance," with "countless numbers of pulleys and straps running in all directions, apparently, to the unskilled eye, in hopeless confusion." In addition to being costly to construct, prone to failure, and a prime cause of accidents, millwork was inefficient. It was not unusual for pulleys and belts to consume a third or more of the power produced by a waterwheel or an engine.

It was into this world that the electric generator made its entry as the third great source of industrial power. Electricity offered a compelling advantage: it didn't require cumbersome millwork. Because every machine could receive power separately, factory owners would gain a new flexibility in their ability to design work flows—and expand their operations. They would no longer be constrained by a complicated array of belts and pulleys that was difficult to modify. Electricity was also cleaner and easier to control than water or steam power.

Adopting electric power was, however, a forbidding proposition. You not only had to sacrifice most of your past investments in water or steam systems and all the attendant millwork, but you had to install a dynamo, run wiring through your plant, and, most daunting of all, retrofit your machines to run on electric motors. It was costly, and because electric power was new and untested it was risky as well. The changeover proceeded slowly at first. In 1900, at the close of the first decade in which electric systems had become a practical alternative for manufacturers, less than 5 percent of the power used in factories came from electricity. But the technological advances of suppliers like General Electric and Westinghouse made electric systems and electric motors ever more affordable and reliable, and the suppliers' intensive marketing programs also sped the adoption of the new technology. Further accelerating the shift was the rapid expansion in the number of skilled electrical engineers, who provided the expertise needed to install and run the new systems. By 1905, a writer for *Engineering* magazine felt comfortable declaring that "no one would now think of planning a new plant with other than electric driving." In short order, electric power had gone from exotic to commonplace.

But one thing didn't change. Factories continued to build their own power-supply systems on their own premises. Few manufacturers considered buying electricity from the small central stations, like Edison's Pearl Street plant, that were popping up across the country. Designed to supply lighting to local homes and shops, the central stations had neither the size nor the skill to serve the needs of big factories. And the factory owners, having always supplied their own power, were loath to entrust such a critical function to an outsider. They knew that a glitch in power supply would bring their operations to a halt—and that a lot of glitches might well mean bankruptcy. "In the early years," as Louis Hunter puts it, "the presumption would be that a manufacturer electrifying his machinery would use his own power plant." That presumption is evident in the statistics. As the new century began, a survey by the Census Bureau found that there were already 50,000 private electric plants in operation, far outstripping the 3,600 central stations.

With the explosion in private systems came a rapid expansion in the industries supplying the components and expertise required to build and operate them. General Electric and Westinghouse became giant companies surrounded by a constellation of smaller suppliers. The vendors and the bankers who backed them had a vested interest in ensuring the continued proliferation of private generating systems. At the time Insull assumed the presidency of Chicago Edison, the idea that a factory owner produced his own power was deeply embedded not only in the heritage of manufacturing itself but in the great and growing electricity industry that served the manufacturers and profited enormously from their business. And at the very center of that industry stood Insull's hero and former boss.

EVEN AS FACTORY owners were rushing to build and expand their own power plants, a pair of technologies were being developed that would render those plants obsolete. In the early 1880s, the English engineer Charles Parson invented a powerful steam turbine that could produce electricity far more efficiently than traditional piston-fired steam engines. Around the same time, the Serbian inventor Nikola Tesla was perfecting a system for distributing electricity as alternating current rather than direct current. In tandem, these breakthroughs fundamentally altered the economics of power supply. The steam turbine allowed central stations to achieve much greater economies of scale in electricity generation and thus push down the cost of producing every kilowatt. Alternating current allowed them to transmit their electricity over great distances and serve a much larger set of customers.

The new alternating-current systems met with considerable resistance at first. Because they operated at much higher voltages than existing systems, they stirred fears about safety among many in the public. Edison, still convinced of the superiority of his own direct-current system, tried to magnify those fears by launching a grisly public relations campaign aimed at having high-power AC systems banned. Teaming up with an electrician named Harold Brown, he helped stage a series of public executions of animals, including dogs, cows, and horses, using the current from AC dynamos. He even convinced the New York legislature to purchase an AC generator—from Westinghouse, which had bought Tesla's patents and become the biggest promoter of AC systems—for use in the electrocution of death-row prisoners. On August 6, 1890, an ax murderer named William Kemmler became the first man to die in New York's new electric chair. Although the next day's newspaper headline—"Kemmler Westinghoused"—must have pleased Edison, his fearmongering failed to halt the spread of the technologically superior AC systems.

While Edison was vainly trying to hold back progress, Insull was moving to capitalize on it. He was the first to realize that, with the new technologies, electricity supply could be consolidated in immense central stations that would be able to meet the demands of even the largest industrial customers. Moreover, utilities' superior economies of scale, combined with their ability to use their capacity much more efficiently by serving many different customers, would allow them to deliver power to factories for a lower price than manufacturers could achieve with their private dynamos. A virtuous circle beckoned: as a utility served more customers, it would become more efficient, allowing it to cut the cost of power further and in turn attract even more customers. "The opportunity to get this large power business was right at my threshold," Insull recalled in his memoirs, "and I knew that unless I built the most economical power station possible, that opportunity would be lost."

Insull rushed to expand the generating capacity of Chicago Edison. When he took over the presidency of the company, on July 1, 1892, it was one of more than twenty small utilities scattered throughout the city, all concentrating on providing electricity for lighting. It operated just two tiny central stations. Insull immediately started work on a much larger plant on Harrison Street near the Chicago River. It was at first outfitted with two 2,400-kilowatt dynamos, but it was designed to accommodate much larger generators. Not long after completing the Harrison Street station, he began planning the construction of an even more ambitious plant, on Fisk Street. He wanted to install in that facility 5,000-kilowatt steam turbines, far larger than any previously built in the country. His supplier, General Electric, balked at the plan and offered to sell him smaller machines instead. But Insull would not be deterred. When he agreed to share the risk of installing the huge turbines, his old employer gave in. It delivered the first 5,000-kilowatt dynamo in 1903. It wouldn't be long before Insull ripped out those machines to install even larger ones. By 1911, the Fisk Street station held ten 12,000-kilowatt turbines.

While he was beefing up his generating capacity, Insull was also buying up his competitors. Less than a year after he joined Chicago Edison, he had already acquired the company's two largest rivals, Chicago Arc Light and Power and Fort Wayne Electric. By 1885, he had gobbled up six more utilities. Soon, he acquired the rest of the central stations operating in Chicago, gaining a monopoly over electricity supply throughout the city. He knew that his success hinged on serving as many customers as possible with his efficient plants. His goal in building a monopoly was not to raise prices but to gain the scale necessary to cut prices drastically—and thus sell even more power to even more customers.

Two other technologies proved crucial to Insull's plans. The first was the rotary converter. Invented in 1888 by Charles Bradley, a former Edison engineer, the rotary converter was a transducer that could turn one form of current into another. As Insull expanded his own plants and bought up others, he found himself with a mix of equipment built for various current standards—direct current, alternating current, and other specialized ones—and operating at different voltages, frequencies, and phases. With rotary converters and other transformers, he was able to meld all his plants into a single system—a much more ambitious version of Edison's one machine—that could be managed centrally. That allowed him to sell electricity for a variety of uses—lights, industrial machinery, even streetcars—through one production operation. The rotary converter made it possible to assemble a universal grid without having to replace all the old equipment.

The second enabling technology was the demand meter, which Insull first saw in operation in 1894 during a vacation in Brighton, England. Unlike traditional meters, which just measured a customer's "load" (the number of kilowatts actually consumed), demand meters also measured the customer's "load factor" (the kilowatts consumed as a percentage of potential peak usage). A customer's peak usage was a vital consideration for a utility because it had to make sure it had enough generating capacity to fulfill the maximum possible demand of its clientele. Customers' peak usage determined a utility's fixed costs—the investments it had to make in building and maintaining its plants and equipment—while their actual usage determined its variable operating costs. The profitability of a utility hinged on its overall load factor, as that determined how efficiently it was using its installed capacity. The higher the load factor, the more money a utility made.

An early example of an information-processing machine, the demand meter opened the way to a revolution in electricity pricing. It allowed utilities to charge each customer two separate fees: a fixed fee reflecting the customer's share of the utility's total fixed costs, and a variable fee reflecting the customer's actual consumption. Instead of charging all buyers the same price for power, utilities could now tailor different pricing schedules to different customers, based on the economics of serving them. Big and relatively efficient users, like factories, could be charged much lower rates than small, less efficient users. By varying prices, moreover, smart utilities could attract a mix of customers whose demand patterns complemented one another—combining heavy nighttime users with heavy daytime users, for instance, or heavy summer users with heavy winter users. Through the careful management of this "diversity factor," as it came to be called, a utility could maximize its load factor and, in turn, its profits.

Insull proved a genius at load balancing, and by the start of the new century he was well on his way to fine-tuning his utility system, at both the technological and the financial level. His achievement in this regard, according to the historian Thomas P. Hughes, was comparable "to the historic managerial contributions made by railway men in the nineteenth century." But Insull still had to convince industrial concerns to close down their private power plants and buy their electricity from his utility.

His first targets were not manufacturers but "traction companies"—the streetcar and elevated railway operators that at the turn of the century were the largest users of electric power in the city. What made these businesses particularly attractive to Insull was their pattern of current consumption. Traction companies required vast amounts of power during the morning and evening rush hours, when workers commuted between their homes and places of work. Their demand perfectly complemented that of home users, whose usage peaked early in the morning and late in the evening, and office users, whose usage peaked during the midday hours. Insull knew that if he could get the traction companies to switch, he would be able to greatly improve his diversity factor. To secure their business, he offered them a rate of less than a penny per kilowatt-hour, far below the prevailing rate of ten cents a kilowatt-hour and much less than they were spending to produce their own power. In 1902, the Lake Street Elevated Railway signed up with Chicago Edison. All the other streetcar and railway operators soon followed suit, dismantling their private generators and plugging in to Insull's network.

With the traction companies' business in hand, Insull launched an aggressive campaign to attract the factories. He set up a downtown "Electric Shop" that included promotional displays of a variety of industrial machines running with electric motors. He placed advertisements in local papers announcing every major new industrial customer he signed up. He used his growing influence to plant laudatory articles in leading trade publications. And he spent heavily on other marketing and sales programs aimed at manufacturers, drilling home the message that he could provide reliable power much less expensively than they could on their own.

It worked. Chicago's manufacturers flocked to Chicago Edison, which Insull soon renamed the Commonwealth Edison Company. In 1908, a reporter for *Electrical World and Engineer* noted that "although isolated plants are still numerous in Chicago, they were never so hard pressed by central station service as now. . . . The Commonwealth Edison Company has among its customers establishments formerly run by some of the largest isolated plants in the city." A year later, the *Electrical Review and Western Electrician* wrote that Insull's customers "now included a large number of great manufacturing and industrial plants." As more manufacturers joined the system, Insull continued to push down prices. Per-capita sales of electricity skyrocketed in Chicago, rising from about 10 kilowatt-hours in 1899 to nearly 450 kilowatt-hours by 1915.

Manufacturers came to find that the benefits of buying electricity from a utility went far beyond cheaper kilowatts. By avoiding the purchase of pricey equipment, they reduced their own fixed costs and freed up capital for more productive purposes. They were also able to trim their corporate staffs, temper the risk of technology obsolescence and malfunction, and relieve their managers of a major distraction. Once unimaginable, the broad adoption of utility power had become inevitable. As other utility operators followed Insull's lead, the transition from private to utility power proceeded rapidly. In 1907, utilities' share of total US electricity production reached 40 percent. By 1920, it had jumped to 70 percent. In 1930, it hit 80 percent. Soon, it was over 90 percent. Only a handful of manufacturers, mainly those running big factories in remote locations, continued to produce their own current.

Thanks to Samuel Insull, the age of the private power plant was over. The utility had triumphed.

CHAPTER 3

Digital Millwork

A S THE TWENTIETH century dawned, companies weren't just retooling their industrial machines to run on the electric current pumped out by utilities. They were also starting to install a very different type of electrical machine, a machine that processed information rather than material and that was operated by office clerks rather than factory hands. The machine, called a punch-card tabulator, had been invented in the early 1880s by an engineer named Herman Hollerith for the purpose of automating the US census. It worked on a simple principle. By punching holes in certain locations on a paper card, you could store information. A single census card, for instance, might hold all the data collected on one family. A hole in one place on the card would indicate that the family had three children, while a hole in another place would indicate that the family lived in an apartment. You could then place the card onto an electrically charged plate in Hollerith's machine and lower a grid of thin metal needles onto it. Wherever a hole was punched, a needle would pass through, completing a circuit and allowing the data on the card to be recorded on a meter. It was a binary system—either a hole was punched in a given location or it wasn't—which anticipated the binary operation of today's digital computers. More than that, though, the way the punch-card tabulator came to be sold and used would set the pattern for the entire modern history of business computing.

The Census Bureau put Hollerith's machine to work in the 1890 census, with great success. The tally proceeded much more quickly than it had in the 1880 round, even though the country's population had grown by about a quarter in the interim. The cost of the census was cut by $5 million—a savings almost ten times greater than the Bureau had expected. Having proved its value in speeding up calculations, the punch-card tabulator attracted the attention of the owners of large businesses like railroads, insurance agencies, banks, and mass-market manufacturers and retailers. As these companies had expanded their operations in the wake of the Industrial Revolution, they'd found it necessary to collect, store, and analyze ever larger amounts of data—on their customers, their finances, their employees, their inventories, and so on. Electrification allowed the companies to grow larger still, further expanding the information they had to process. This intellectual work became as important, and often as arduous, as the physical labor of manufacturing products and delivering services. Hollerith's tabulator allowed big businesses to process information far more quickly, with fewer people and greater accuracy, than was possible before.

Seeing the commercial potential of his invention, Hollerith established the Tabulating Machine Company to sell tabulators to businesses. The company grew rapidly, introducing a series of related products such as alphabetic tabulators, card sorters, card duplicators, and printers and selling them to an ever broader clientele. In 1911, Hollerith's firm merged with the Computer–Tabulating–Recording Company, an even larger supplier of business machines. A talented young manager named Thomas J. Watson was brought in to run the business. Thirteen years later, the ambitious Watson changed the company's name to the more impressive sounding International Business Machines Corporation. Other companies, like Burroughs and Remington Rand in the United States and Bull in Europe, also rushed into the burgeoning punch-card market, competing with Watson's IBM.

The information technology industry had been born.

The spread of punch-card technology accelerated as standards were established for the design of the cards and the operation of the equipment and as technical advances and competition pushed down prices. Within a couple of decades, most large companies had set up punch-card rooms for the various machines they used to sort, tabulate, and store financial and other business information. They invested a great deal of capital in the machinery, and they hired clerks and technical specialists to operate and maintain it. They also developed close ties with the suppliers of the systems. "Processing information gathered into a deck of cards was entrenched into business practices by the mid-1930s," computer historian Paul Ceruzzi writes, "and reinforced by the deep penetration of the punched-card equipment salesmen into the accounting offices of their customers."

Even as companies were dismantling their power-generation departments, in other words, they were establishing new departments dedicated to the fledgling technology of automated data processing. During the second half of the century, these departments would grow dramatically as electronic digital computers replaced punch-card machines. Most big companies would find themselves building ever more elaborate installations of computing hardware and software, spending tens or even hundreds of millions of dollars a year on their in-house computing operations, and relying ever more heavily on IT vendors and consultants to keep their systems running. As the manipulation of symbols—words, numbers, images—supplanted the manipulation of materials as the focus of business in the developed world, the private power-generating plant of the nineteenth century found its echo in the private data-processing plant of the twentieth. And just as before, companies came to assume that there was no alternative—that running a complex computing operation was an intrinsic part of doing business.

ALTHOUGH IT NOW seems inevitable that the computer would become the mainstay of modern business, there was originally much skepticism about the machine's usefulness. When the first true commercial computer, the UNIVAC, was being built in the 1940s, few people believed it had much of a future in the corporate world. At the time, it was hard to imagine that many companies would have a need for the kind of intensive mathematical calculations that an electronic computer could churn out. The old punch-card tabulators seemed more than sufficient for handling transactions and keeping accounts. Howard Aiken, a distinguished Harvard mathematician and a member of the US government's National Research Council, dismissed as "foolishness" the idea that there would be a big market for computers. He believed that the country would need no more than a half dozen of them, mainly for military and scientific research. Even Thomas Watson is reputed to have said, in 1943, "I think there is a world market for about five computers."

But the designers of the UNIVAC, two University of Pennsylvania professors named J. Presper Eckert and John Mauchly, saw things differently. They realized that because an electronic computer would be able to store its operating instructions in its own memory, it could be programmed to perform many functions. It wouldn't just be a glorified calculator, limited to preset mathematical routines. It would become a general purpose technology, a machine-of-all-trades, that companies could apply not just to everyday accounting chores but to innumerable managerial and analytical tasks. In a 1948 memo, Mauchly listed nearly two dozen companies, government agencies, and universities that he believed would be able to put the UNIVAC to good use. As it turned out, the market proved a good deal larger than even he expected.

Leading the way in adopting the powerful new machines was, once again, the Census Bureau. On March 31, 1951, it purchased the first UNIVAC, installing it a year later in its headquarters in Washington DC. By the end of 1954, Eckert and Mauchly's computers were running in the offices of ten private corporations, including General Electric, US Steel, Du Pont, Metropolitan Life, Westinghouse, and Consolidated Edison, the descendant of Thomas Edison's Edison Electric Illuminating Company. UNIVACs performed all the jobs that punch-card systems did—billing, payroll management, cost accounting—but they were also used for more complicated tasks like sales forecasting, factory scheduling, and inventory management. In short order, skepticism about the role of computers in business gave way to rampant enthusiasm. "The Utopia of automatic production is inherently plausible," the *Harvard Business Review* proclaimed in the summer of 1954.

The enthusiasm spread to the manufacturers of business machines, who saw in computing an expansive and lucrative new market. Soon after the UNIVAC appeared, IBM introduced its own line of mainframe computers, the 701 series, and by 1960 Honeywell, General Electric, RCA, NCR, Burroughs, and AT&T's Western Electric division were all in competition to sell computer gear. An entirely new industry—software programming—also began to take shape. About forty small software companies, with names like Computer Sciences Corporation, Computer Usage Company, and Computer Applications Inc., were founded during the late 1950s to write programs for mainframes.

It wasn't long before businesses were competing not only on the quality of their products but on the capabilities of their hardware and software. Once one company introduced a new system to automate an activity, other companies, fearful of being put at a disadvantage, followed suit. The first battles in what would become a worldwide information-technology arms race took place in the airline business. In 1959, Cyrus Rowlett Smith, the president of American Airlines, launched an ambitious project to build a system that would automate the making of flight reservations and the issuing of tickets, two labor-intensive processes that lay at the heart of the business. Built by 200 technicians over the course of more than five years, the system, called Sabre, incorporated two of IBM's most powerful mainframe computers as well as sixteen data-storage devices and more than 1,000 terminals for ticket agents. In addition to assembling the hardware, the project entailed the writing of a million lines of software code. When the system went into full operation at the end of 1965, it was able to process 40,000 reservations and 20,000 ticket sales a day—an astonishing feat at the time.

Sabre provided as great an advantage to American Airlines as Burden's waterwheel had provided to his ironworks. American was able to operate with fewer employees and higher productivity than other airlines, which continued to process reservations by hand. It also enjoyed big benefits in customer service, as it was able to respond to travelers' requests and inquiries far more quickly than its rivals could. It gained an intelligence edge as well, as it could monitor demand for different routes and adjust ticket prices with great precision. Building and running computer systems had become as important to the success of American Airlines as flying planes and pampering passengers. In the years that followed, all the other major airlines, including Pan American, Delta, and United, built similar systems. They saw that they had no choice if they wanted to remain competitive. Not surprisingly, they found eager partners in computer vendors like IBM, Sperry Rand, and Burroughs, which earned big profits by replicating similar systems in one company after another.

Bank of America initiated a similar cycle of copycat investments in the banking industry when, in 1960, it unveiled its ground-breaking Electronic Recording Machine Accounting computer in a televised extravaganza hosted by Ronald Reagan. Within two years, the bank had thirty-two ERMA computers up and running, processing nearly five million checking and savings accounts that formerly had to be updated manually. The computers' ability to handle transactions with unprecedented speed and accuracy forced all major financial institutions to follow in Bank of America's foot-steps. The same phenomenon was soon playing out in every other industry, as companies matched one another's investments in the latest computer gear.

But the mainframe era proved to be only the very start of the business world's great computer shopping spree. At the close of the 1960s, the average American company devoted less than 10 percent of its capital equipment budget to information technology. Thirty years later, that percentage had increased more than fourfold, to 45 percent, according to Commerce Department statistics. By 2000, in other words, the average US company was investing almost as much cash into computer systems as into all other types of equip-ment combined. Spending on software alone increased more than a hundred-fold during this period, from $1 billion in 1970 to $138 billion in 2000. The rest of the developed world saw a similar burst in investment, as global IT expenditures leapt from less than $100 billion a year in the early 1970s to more than $1 trillion a year by the early 2000s.

What happened during those thirty years? Businesses changed, and computers changed. As the economy became weighted more toward services and less toward manufacturing, investment shifted away from industrial machinery and into information technology. At the same time, computers themselves became smaller, cheaper, easier to program, and more powerful, dramatically expanding the range of tasks they could be applied to. Most important of all, computers became personal—they turned into commonplace tools that nearly all office employees could use.

During the mainframe era, by contrast, computers were institutional machines. Because it was so expensive to buy or lease a mainframe—the rent on a typical IBM computer was about $30,000 a month in the mid-1960s—a company had to keep the machine in constant use if it was to justify the expense. That meant that individual employees almost never had direct access to a computer. Like the punch-card tabulators that preceded them, mainframes and related gear were secluded in special rooms and operated by dedicated staffs of white-suited specialists—a "priesthood of technicians," in Ceruzzi's words. To use one of the machines, an employee had to store the program he wanted to run, along with all the necessary data, on a loop of tape or a deck of cards, and then put his job, or "batch," into a queue along with the jobs of his colleagues. The operators of the mainframe would run one batch after another, bundling the results into printouts that employees would then pick up and review. If an employee found an error, he would have to resubmit his batch and go through the entire cycle again.

Mainframe batch-processing had one big advantage: it ensured that a computer was used efficiently. No machine sat idle, at least not for long. The typical corporate mainframe operated at more than 90 percent of its full capacity. But batch-processing had an even larger disadvantage: it rendered computing impersonal. The organizational and technological barriers that stood between the employee and the machine stifled experimentation and constrained the application of computing power, while the delay in receiving results prevented computers from being used to support the many small, everyday decisions necessary to keep a company running.

This shortcoming didn't last long. As technological innovation sped forward through the 1960s and 1970s, computers shrank in size and price. Tiny transistors replaced bulky vacuum tubes, and cheap standardized components replaced expensive custom parts, giving rise to relatively inexpensive minicomputers that could fit beside a desk. Minicomputers didn't supplant mainframes; they supplemented the bigger, more powerful machines. But they greatly expanded businesses' use of computers. Because minicomputers could be connected to desktop terminals, they allowed ordinary employees to tap directly into the power of a computer to do a wide range of jobs, from analyzing business investments to designing new products to scheduling assembly lines to writing letters and reports. The languages for writing software also became much simpler during this time. Programmers could write code using basic English words and syntax rather than long lines of numbers. That greatly expanded the computer programming industry, bringing a big jump in the number of programmers and in the kinds of applications they developed. By the early 1970s, a company could buy a minicomputer for less than $10,000 and quickly program it to carry out a specialized task.

The minicomputer business flourished, propelling companies like Digital Equipment, Wang, and Apollo into the front ranks of the computer industry. But its heyday proved brief. The mini turned out to be a transitional machine. Breakthroughs in the design of integrated circuits, particularly the invention of the microprocessor by Intel engineers in 1971, led to the introduction and rapid proliferation of an entirely new kind of machine—the micro, or personal, computer—that was even smaller, even cheaper, and even easier to use than the mini. The arrival of the microcomputer would quickly upend the industry, ushering in a new era in business computing.

AS WITH MAINFRAMES, experts didn't at first see much potential for personal computers in business. This time, though, the doubts were of a very different sort. Where mainframes were viewed as too powerful for commercial applications, PCs were seen as too weak. They were dismissed as frivolous gadgets, toys for geeky hobbyists. The dominant computer companies of the day, from IBM to Digital, paid the quirky new machines little mind. It took a college dropout named Bill Gates—himself a geeky hobbyist—to see the potential of personal computers in business. In 1975, Gates and his high-school buddy Paul Allen founded a little company named Micro-Soft to write software for the newly invented PC. Gates soon realized that the machine would not only find a place inside business but that, because of its versatility and low cost, it would supplant the mainframe as the center of corporate computing. Any company able to gain control over the PC's operating system, and the virtual desktop it created, would become the most powerful in the computing business. Gates's insight would make Microsoft—as he renamed the company—the dominant player in the IT industry and bring Gates himself unimaginable wealth.

The PC democratized computing. It liberated the computer from corporate data centers and IT departments and turned it into a universal business tool. In the process, it also changed the way companies organized their computing assets and operations. The personal computers sitting on office workers' desks soon came to be wired together into networks to allow employees to exchange files and share printers. The old centralized mainframe rooms didn't disappear. They were transformed into a new kind of data center. The engine rooms of modern businesses, these centers contained the storage systems that held companies' most important data as well as the powerful server computers that ran the applications used to manage their finances and operations. Individual employees could use their PCs to run their own personal programs, such as Microsoft Word and Excel, but they could also use them to tap into the programs and files on the central servers. Because the PCs acted as "clients" of the shared servers, this setup came to be known as "client–server computing." It became the defining model of corporate computing in the PC age, the model that has remained dominant to the present day.

Client–server computing has turned out to be the mirror image of mainframe computing. It has made computing personal, but it has also made it woefully inefficient. Corporate computer systems and networks, the digital millwork of the modern company, have become steadily more complex as their applications have multiplied. One of the main reasons for the complexity is the historical lack of standards in computing hardware and software. Vendors have tended to promote their own proprietary products, which by design don't mesh well with competitors' gear. As a result, corporate software programs have generally been written to run on a particular operating system, a particular microchip, a particular database, and a particular hardware setup. Unlike the multipurpose mainframes, most server computers have had to be used as single-purpose machines, dedicated to running just one software application or one database. Whenever a company buys or writes a new application, it has to purchase and install another set of dedicated computers. Each of these computers, moreover, has to be configured to handle the peak theoretical demand for the application it runs—even if the peak load is rarely or ever reached.

The proliferation of single-purpose systems has resulted in

extraordinarily low levels of capacity utilization. One recent study of six corporate data centers revealed that most of their 1,000 servers were using less than a quarter of their available processing power. Other studies indicate that data storage systems are almost equally underused, with capacity utilization averaging between 25 and 50 percent. Before the PC age, data-processing professionals viewed the conservation of computing resources as not just an economic imperative but an ethical one. "To waste a CPU cycle or a byte of memory was an embarrassing lapse," recalls the science writer Brian Hayes. "To clobber a small problem with a big computer was considered tasteless, and unsporting, like trout fishing with dynamite." The client–server model killed the conservation ethic. Profligacy replaced frugality as the defining characteristic of business computing.

The complexity and inefficiency of the client–server model have fed on themselves over the last quarter century. As companies continue to add more applications, they have to expand their data centers, install new machines, reprogram old ones, and hire ever larger numbers of technicians to keep everything running. When you also take into account that businesses have to buy backup equipment in case a server or storage system fails, you realize that, as studies indicate, most of the many trillions of dollars that companies have invested into information technology have gone to waste.

And there are other costs as well. As data centers have expanded and become more densely packed with computers, electricity consumption has skyrocketed. According to a December 2005 study by the Department of Energy's Lawrence Berkeley National Laboratory, a modern corporate data center "can use up to 100 times as much energy per square foot as a typical office building." The researchers found that a company can spend upwards of $1 million per month on the electricity required to run a single big data center.

And the electric bill continues to mount rapidly as servers proliferate and computer chips become more powerful and power-hungry.
Luiz André Barroso, a computer engineer with Google, concludes
that, barring substantial improvements in the efficiency of computers, "over the next few years, power costs could easily overtake
hardware costs, possibly by a large margin."

The waste inherent in client–server computing is onerous for individual companies. But the picture gets worse—much worse—when
you look at entire industries. Most of the software and almost all of
the hardware that companies use today are essentially the same as
the hardware and software their competitors use. Computers, storage systems, networking gear, and most widely used applications
have all become commodities from the standpoint of the businesses
that buy them. They don't distinguish one company from the next.
The same goes for the employees who staff IT departments. Most
perform routine maintenance chores—exactly the same tasks that
their counterparts in other companies carry out. The replication
of tens of thousands of independent data centers, all using similar
hardware, running similar software, and employing similar kinds
of workers, has imposed severe penalties on the economy. It has led
to the overbuilding of IT assets in almost every sector of industry,
dampening the productivity gains that can spring from computer
automation.

The leading IT vendors have ridden the investment wave to
become some of the world's fastest-growing and most profitable
businesses. Bill Gates's company is a perfect case in point. Almost
every company of any size today buys copies of Microsoft Windows
and Microsoft Office for all its white-collar workers, installing the
software individually on every PC and upgrading the programs routinely. Most also run at least some of their servers on a version of the
Windows operating system and install other expensive Microsoft

programs in their data centers, such as the Exchange software used to manage email systems. In the three decades since its founding, Microsoft grew to have annual sales of nearly $50 billion, annual profits of more than $12 billion, and more than $30 billion of cash in the bank. And Microsoft has plenty of company, from other software makers like Oracle and SAP to server suppliers like IBM and Hewlett–Packard to PC vendors like Dell to the hundreds of consulting firms that feed off the complexity of modern business computing. They've all happily played the role of weapons suppliers in the IT arms race.

WHY HAS COMPUTING progressed in such a seemingly dysfunctional way? Why has the personalization of computers been accompanied by such complexity and waste? The reason is fairly simple. It comes down to two laws. The first and most famous was formulated in 1965 by the brilliant Intel engineer Gordon Moore. Moore's Law says that the power of microprocessors doubles every year or two. The second was proposed in the 1990s by Moore's equally distinguished colleague Andy Grove. Grove's Law says that telecommunications bandwidth doubles only every century. Grove intended his "law" more as a criticism of what he considered a moribund telephone industry than as a statement of technological fact, but it nevertheless expresses a basic truth: throughout the history of computing, processing power has expanded far more rapidly than the capacity of communication networks. This discrepancy has meant that a company can only reap the benefits of advanced computers if it installs them in its own offices and hooks them into its own local network. As with electricity in the time of direct-current systems, there's been no practical way to transport computing power efficiently over great distances.

As Grove's observation reveals, the scarcity of communications bandwidth has long been recognized as a barrier to effective and efficient computing. It has always been understood that, in theory, computing power, like electric power, could be provided over a grid from large-scale utilities—and that such centralized dynamos would be able to operate much more efficiently and flexibly than scattered, private data centers. Back in 1961, when computer scientists were just beginning to figure out how to get computers to talk to each other, one expert in the fledgling art of networking, John McCarthy, predicted that "computing may someday be organized as a public utility, just as the telephone system is organized as a public utility." Every advance in networking brought a new wave of entrepreneurs hoping to turn utility computing into a big business. In the mainframe era, time-sharing companies set up central computers and rented them out to other businesses, allowing direct connections over phone lines. In the 1970s, companies like Automated Data Processing began to provide some routine computing jobs—payroll processing, notably—as for-fee services. And in the 1990s, a slew of "application service providers" emerged, with considerable venture-capital backing, in hopes of providing businesses with software programs over the Internet.

But all these attempts at utility computing were either doomed or hamstrung by the lack of sufficient bandwidth. Even during the late 1990s, as the telecommunications industry rushed to modernize its networks, broadband capacity was still neither inexpensive nor plentiful enough for utilities to deliver computing services with the speed and reliability that businesses enjoyed with their local machines. And so companies continued to entangle themselves in their digital millwork, accepting complexity, inefficiency, and waste as the costs of automating their operations.

But now, at last, that's changing. The network barrier has, in just the last few years, begun to collapse. Thanks to all the fiber-optic cable laid by communications companies during the dotcom boom—enough, according to one estimate, to circle the globe more than 11,000 times—Internet bandwidth has become abundant and abundantly cheap. Grove's Law has been repealed. And that, when it comes to computing at least, changes everything. Now that data can stream through the Internet at the speed of light, the full power of computers can finally be delivered to users from afar. It doesn't matter much whether the server computer running your program is in the data center down the hall or in somebody else's data center on the other side of the country. All the machines are now connected and shared—they're one machine. As Google's chief executive, Eric Schmidt, predicted way back in 1993, when he was the chief technology officer with Sun Microsystems, "When the network becomes as fast as the processor, the computer hollows out and spreads across the network."

What the fiber-optic Internet does for computing is exactly what the alternating-current network did for electricity: it makes the location of the equipment unimportant to the user. But it does more than that. Because the Internet has been designed to accommodate any type of computer and any form of digital information, it also plays the role of Insull's rotary converter: it allows disparate and formerly incompatible machines to operate together as a single system. It creates harmony out of cacophony. By providing a universal medium for data transmission and translation, the Net is spurring the creation of centralized computing plants that can serve thousands or millions of customers simultaneously. What companies used to have no choice but to supply themselves, they can now purchase as a service for a simple fee. And that means they can finally free themselves from their digital millwork.

It will take many years for the utility computing system to mature. Like Edison and Insull before them, the pioneers of the new industry will face difficult business and technical challenges. They'll need to figure out the best ways to meter and set prices for different kinds of services. They'll need to become more adept at balancing loads and managing diversity factors as demand grows. They'll need to work with governments to establish effective regulatory regimes. They'll need to achieve new levels of security, reliability, and efficiency. Most daunting of all, they'll need to convince big companies to give up control over their private systems and begin to dismantle the data centers into which they've plowed so much money. But these challenges will be met just as they were met before. The economics of computing have changed, and it's the new economics that are now guiding progress. The PC age is giving way to a new era: the utility age.

CHAPTER 4

Goodbye, Mr. Gates

"THE NEXT SEA change is upon us." Those words appeared in an extraordinary memorandum that Bill Gates sent to Microsoft's top managers and engineers on October 30, 2005. Belying its bland title, "Internet Software Services," the memo was intended to sound an alarm, to warn the company that the rise of utility computing threatened to destroy its traditional business. What had always been the linchpin of Microsoft's success—its dominance over the PC desktop—was fading in importance. Software, Gates told his troops, was no longer something people had to install on their own computers. It was turning into a utility service supplied over the Internet. "The broad and rich foundation of the internet will unleash a 'services wave' of applications and experiences available instantly," he wrote, using the jargon of the technocrat. "Services designed to scale to tens or hundreds of millions [of users] will dramatically change the nature and cost of solutions deliverable to enterprises or small businesses." This new wave, he concluded, "will be very disruptive."

It's not hard to understand what spurred the memo. As Gates composed it in his office at Microsoft's headquarters in Redmond, Washington, his fears about his company's future were taking concrete shape just a couple of hundred miles away, in a sleepy town named The Dalles in northern Oregon. Earlier that year, a mysterious company, known only as Design LLC, had quietly entered into negotiations with local officials to buy a 30-acre parcel of land owned by a government agency. The land, part of a large industrial park, lay along the banks of the Columbia River. Hoping to keep the negotiations under wraps, the company required town officials, including the city manager, to sign confidentiality agreements. But details of the shadowy deal soon began to leak out. Design LLC, it turned out, was a front. The corporation that was actually interested in purchasing the site was none other than Google, the dominant Internet search company that was rapidly becoming Microsoft's most dangerous adversary.

In February 2005, Google closed the deal, buying the parcel for just under $2 million after the town agreed to its demand for tax breaks. A few hundred construction workers and a fleet of backhoes, dump trucks, and cement mixers moved in to begin work on "Project 2," as the job was code-named. The undertaking required what Eric Schmidt would later call a "massive investment." As work progressed over the course of the year, the extent of the facility became apparent. Two large, windowless warehouses, each about the size of a football field, dominated the site. Above them rose four cooling towers, lending the complex an ominous look. An article in the *International Herald Tribune* portrayed it as "looming like an information-age nuclear plant."

It was an apt description. What Google was building was a vast data-processing plant, by all accounts the largest and most sophisticated on the planet. Designed to house tens or even hundreds of thousands of computers, all working together as a single machine, it was, indeed, the information-processing equivalent of a nuclear power plant, a data dynamo of unprecedented power. We may in the future look back on Google's The Dalles plant as just an early and relatively primitive example of a central computing station, similar to the way we now see Samuel Insull's Fisk Street electric plant, but today it represents the state of the art in utility computing. And as Gates no doubt understood in October 2005, it symbolizes the upheaval that is reshaping the computer industry—and throwing Microsoft's future into doubt.

By the time Google began work on the plant, the company had already set up dozens of other "server farms" in covert locations around the world. Altogether they held as many as a half million computers. But its need for raw computing power had continued to soar. The Dalles, which one Oregon newspaper described as "a hamburger-and-gas pit stop between Portland and Pendleton," turned out to be the perfect site for the company's most ambitious center yet. The town's remoteness would make it easier for Google to keep the facility secure—and harder for its employees to be lured away by competitors. More important, the town had ready access to the two resources most critical to the data center's efficient operation: cheap electricity and plentiful bandwidth. Google would be able to power its computers with the electricity produced by the many hydroelectric dams along the Columbia, particularly the nearby The Dalles dam with its 1.8-gigawatt generating station. It would also be able to temper its demand for electricity by tapping the river's icy waters to help cool its machines. As for bandwidth, the town had invested in building a large fiber-optic data network with a direct link to an international Internet hub in nearby Harbour Pointe, Washington. The network provided the rich connection to the Internet that Google needed to deliver its services to the world's Web surfers.

Google's data centers have been designed by some of the best minds in computer science. Like Edison's one machine, they operate as a finely tuned system—what the legendary computer designer Danny Hillis calls "the biggest computer in the world"—with each component meticulously engineered to work seamlessly with every other. Each center contains one or more "clusters" of custom-built server computers. The servers are little more than homemade PCs constructed of cheap commodity microprocessors and hard drives that Google purchases in bulk directly from their manufacturers. Rather than being hard-wired together inside boxes, the components are simply attached to tall metal racks with Velcro, making it easy to swap them out should they fail. Each computer receives its electricity through a power-supply unit invented by Google engineers to minimize energy consumption, and the machines run a version of the free Linux operating system tweaked by Google's coders. The company even owns much of the fiber-optic cabling that links its centers together, allowing it to precisely control the flow of data among them and between them and the public Internet.

The most important element of the system, the glue that holds it all together, is the proprietary software that Google has written to coordinate, at one level, all the servers in a cluster and, at a higher level, all the clusters in all the firm's centers. Although the company has been extremely secretive about its technology, we know in general terms how the software works in performing Web searches. In its database, Google maintains a copy of virtually the entire Internet, gathered and continually updated by "spidering" software that creeps through the Web, link by link, scanning the contents of the billions of pages it discovers. A set of secret algorithms analyzes all the pages to create a comprehensive index of the Web, with every page ranked according to its relevance to particular keywords. The index is then replicated in each cluster. When a person enters a keyword into Google's search engine, the software routes the search to one of the clusters, where it is reviewed simultaneously by hundreds or thousands of servers. Because each server only has to compare the keyword to a small portion of the entire index—what Google calls an "index shard"—this kind of "parallel processing" proceeds much faster than if a single computer had to compare the keyword to the entire index. The software collects and synthesizes all the responses from the servers, ranks the matching pages according to their relevance, and sends the list of results back to the searcher's computer.

Although a typical search requires, according to Google engineers, "tens of billions of [microprocessor] cycles" and the reading of "hundreds of megabytes of data," the whole process is completed in a fraction of a second. It's the coordination software, acting as a kind of traffic cop, that makes sure the processing load is balanced among all the clusters and the individual servers. It keeps the entire system working as speedily and efficiently as possible. And when a computer or one of its subcomponents fails, the software simply routes work around it. Because the system is constructed of so many thousands of components, none of which is essential, it's failsafe. It can't break down.

No corporate computing system, not even the ones operated by very large businesses, can match the efficiency, speed, and flexibility of Google's system. One analyst estimates that Google can carry out a computing task for one-tenth of what it would cost a typical company. That's why the Google dynamo makes Bill Gates and other technology executives so nervous. It encapsulates the full disruptive potential of utility computing. If companies can rely on central stations like Google's to fulfill all or most of their computing requirements, they'll be able to slash the money they spend on their own hardware and software—and all the dollars saved are ones that would have gone into the coffers of Microsoft and the other tech giants. The traditional suppliers can't even take comfort in the hope that they'll be able to supply their products to the new utilities. Google, after all, builds its own computers and uses free open-source software. It has little need for the old vendors.

As Google has expanded its computing utility, it has been able to rapidly introduce new services as well as acquire ones developed by other companies. Many of those services, from the Google Earth mapping tool to the YouTube video-hosting site to the Blogger weblog publisher, are aimed mainly at consumers. But Google has also begun to muscle into the business market. It has launched a popular package of services, called Google Apps, that competes directly with one of Microsoft's biggest moneymakers, the Office suite. Google Apps includes word processing, spreadsheets, email, calendars, instant messaging, and Web-site design and hosting. It costs just $50 per employee per year—and a basic version, with advertisements, can be had for free. Using the programs requires nothing more than a cheap PC and a browser. Already, many small companies can satisfy most of their everyday computing needs with the software running in Google's data centers. As Google continues to grow—and in 2007 it announced plans to build new centers in North Carolina, South Carolina, Oklahoma, and Iowa—many more businesses will fall into that category.

WHILE GOOGLE MAY at the moment be the biggest force in utility computing, it is by no means the only company pioneering this new business. Other, more specialized "software as a service" firms are offering traditional business programs—for managing finances, say, or coordinating sales and marketing—over the Internet. And they're making big inroads into the corporate market.

The leader in this field is a rapidly growing San Francisco outfit called Salesforce.com. Its founder, Marc Benioff, would never be mistaken for Samuel Insull—a burly extrovert with shaggy hair and a scraggly beard, Benioff has what's been called "the biggest mouth in Silicon Valley"—but his story is not all that different from Insull's. He spent thirteen years working at the tech giant Oracle, where he found a mentor in the company's charismatic CEO, Larry Ellison. Oracle's relational-database software, like Microsoft's operating system and business applications, had in the 1980s and 1990s become a common component of client–server systems, spurring the company's rise to the upper reaches of the computer business and making Ellison a multibillionaire. Benioff thrived at the company, becoming one of its top managers.

But just as Insull became disenchanted with Edison and the business strategy he was pursuing at General Electric, Benioff grew restive at Oracle during the late 1990s. Although the company was minting money at the time, thanks to the euphoria over "e-commerce," Benioff came to believe that the entire corporate software industry was doomed. Its traditional way of doing business—installing big, complex programs on clients' computers and then charging them high licensing and maintenance fees year after year—could not be sustained much longer. The growth of the Internet, Benioff saw, would soon allow companies to avoid the headaches and high costs of owning and running their own applications.

In 1999, proclaiming the imminent "end of software" to any reporter who would listen, Benioff left Oracle and, together with a gifted software engineer named Parker Harris, launched Sales force.com. The company would provide a common type of business software called customer relationship management, or CRM, which helps corporate sales forces keep track of their accounts. CRM was a large and lucrative segment of the software industry, but it had also come to symbolize all the industry's flaws. CRM systems were difficult to install, complicated to use, and they often cost hundreds of thousands of dollars. The companies that bought the systems rarely earned back a return on their investment. But software vendors, particularly the leading CRM company, Siebel Systems, profited mightily.

What Benioff was offering companies was altogether different. They wouldn't have to buy software licenses or maintenance contracts. They wouldn't need to invest in any new servers or other equipment. They wouldn't have to hire consultants to "integrate" their systems. Their marketers and salespeople could simply launch their Web browsers, click over to the Salesforce site, and start working. All the software code, and all their data, resided on Salesforce's computers. And when the program needed to be upgraded, the new version simply appeared. The price was startlingly low compared to traditional CRM systems—$50 per user per month—and companies could test-drive the system for free to make sure it fit their needs.

Benioff faced a wall of skepticism when he launched his firm. To buyers who had been burned by the software industry's grandiose promises, his pitch seemed too good to be true. And prospective customers had practical concerns as well: How reliable would the service be? Would it disappear, along with their data, should Salesforce go out of business? Would it be fast enough? Would they be able to customize it? What would happen if they lost their Internet connection? And what about the security of their information? If it was floating around in a system shared by many other companies, including competitors, could it be compromised? The disclosure of a company's data on its customers and their purchases could, after all, be devastating.

But Salesforce was able to address the concerns. Its system turned out to be at least as reliable as most corporate systems. Response times, measured in milliseconds, were often indistinguishable from those achieved by client–server systems. Customers could easily tweak the way information was laid out, and they could even write custom code that would run on Salesforce's computers. By saving, or "caching," some data on users' hard drives, Salesforce ensured that people could continue to work even if they were temporarily disconnected from the Net—during a plane flight, say. And by using cutting-edge data encryption techniques, Salesforce was able to maintain the integrity of each customer's information.

As it turned out, the idea of software-as-a-service caught on even more quickly than Benioff expected. In 2002, the firm's sales hit $50 million. Just five years later, they had jumped tenfold, to $500 million. It wasn't just small companies that were buying its service, though they had constituted the bulk of the earliest subscribers. Big companies like SunTrust, Merrill Lynch, Dow Jones, and Perkin-Elmer had also begun to sign up, often abandoning their old in-house systems in the process. Benioff's audacious gamble, like Insull's a century earlier, had panned out. As for the once mighty Siebel Systems, it had gone out of business as a stand-alone company. After suffering a string of deep losses in the early years of the decade, it was bought up in early 2006 by Benioff's old company, Oracle.

In Salesforce's wake, hundreds of startups have launched software-as-a-service businesses. Some, like RightNow Technologies, compete with Salesforce in the CRM market. Others apply the new model to different kinds of popular business programs. Employease offers a service for managing personnel. LeanLogistics has one for scheduling transportation. Oco provides a "business intelligence" service, allowing executives to analyze corporate information and create planning reports. Digital Insight supplies a range of services to banks. Workday and NetSuite offer online versions of complete "enterprise resource planning" packages—the kind of complex systems for managing finances and operations that can cost millions of dollars when purchased from traditional suppliers like SAP.

In fact, nearly every traditional business application now has an equivalent offered over the Internet, and companies are eagerly embracing the new services. A survey by the management consultancy McKinsey & Company found that 61 percent of large companies planned to use at least one utility software service during 2007, a dramatic increase over the 38 percent who had had similar plans just a year earlier. Gartner, a big IT research house, reports that software-as-a-service sales are booming and will by 2011 account for 25 percent of the business software market, up from a mere 5 percent in 2005.

OTHER COMPANIES ARE taking a different approach to utility computing. Instead of distributing software programs, they're selling hardware itself as a service. They're setting up big data centers and then letting customers tap directly into them over the Net. Each customer is free to choose which programs it wants to run and which types of data it wants to store.

As surprising as it may seem, one of the early leaders in the hardware-as-a-service field is the online retailing giant Amazon. com. Early in its history, Amazon began providing simple utility services to its retailing partners. Companies and individuals could become "Amazon affiliates," giving them the right to market Amazon's products through their own sites (in return for a small cut of any resulting sales). Amazon gave these affiliates a set of software tools that allowed them to connect, via the Internet, with the Amazon databases that stored product descriptions and photographs, customer reviews, prices, and other information needed for effective merchandising. It later expanded these services to allow other retailers to sell their own merchandise through the Amazon site. The retailing services proved so popular, and spurred so much creativity on the part of affiliates and other partners, that in 2002 the company set up a separate subsidiary, Amazon Web Services, to oversee and expand this part of its business. The unit soon made an audacious decision: it wouldn't just provide access to the retailing information in its system; it would open up the system itself. It would let any company keep its information and run its software on Amazon's computers.

Amazon launched its first utility computing service in March 2006. Called Simple Storage Solution, or S3, it allows customers to store data on Amazon's system for a few cents per gigabyte per month. In July of the same year, the company introduced Simple Queue Service, which customers can use to exchange digital messages between various software applications to coordinate their operation. The most ambitious of the services was announced a month later. Called Amazon Elastic Compute Cloud, or EC2, it lets a customer run software programs directly on Amazon's system—to use Amazon's computers as if they were its own. The cost is just ten cents per computer per hour.

Using the services, a company can run a Web site or a corporate software application, or even operate an entire Internet business, without having to invest in any server computers, storage systems, or associated software. In fact, there are no upfront costs whatsoever—a company only pays for the capacity it uses when it uses it. And what it's renting isn't any ordinary computing system. It's a state-of-the-art system designed for modern Internet computing, offering high reliability, quick response times, and the flexibility to handle big fluctuations in traffic. Any company, even a one-person home business, can piggyback on a computing operation that Amazon took years to assemble and fine-tune.

It's no surprise that most of Amazon's early clients have been smaller firms—ones that can't afford to build such an advanced system on their own. Suddenly, these companies, which have long lagged behind big businesses in reaping the benefits of computer automation, can catch up. Amazon's utility levels the playing field. The Internet photo-hosting site SmugMug is a good example. As its site became more popular, it was flooded by the large image files uploaded by its users. The number of files jumped to half a billion, and demand showed no signs of slackening. Instead of investing in expensive storage gear, it simply enrolled in the S3 service, using Amazon's mammoth storage capacity as the data warehouse for its own site. SmugMug estimates that it saved $500,000 in equipment costs as a result—and was able to avoid hiring more staff and renting more office space. The utility service, says the company's CEO, Don MacAskill, "makes it possible for SmugMug to compete with huge, deep-pocketed companies without having to raise massive amounts of cash for hardware." The founder of another startup using S3 says, "It's like having Amazon engineers working for us."

Amazon wasn't only thinking about making its customers' lives easier when it went into the utility computing business. Like every other sizable company today, it had been forced to purchase far more computing and storage capacity than it would ever use. It had to build its system to handle the largest amount of traffic that its online store might experience, and then it had to add an extra dollop of capacity for safety's sake. To put it another way, Amazon had to construct its system to be large enough to accommodate the burst of shopping during the week after Thanksgiving—even though that week comes around only once a year. Most of the system's capacity went unused most of the time. In fact, the company's founder and chief executive, Jeff Bezos, confessed in a 2006 interview that "there are times when we're using less than 10 percent of capacity." By renting out the system to others, Amazon can boost its capacity utilization and slash the overall price of computing, not just for its clients but for itself. As with electric utilities, the more clients Amazon serves, the more balanced its load becomes, pushing its overall utilization rates even higher. Amazon's decision to go into utility computing—an unlikely choice for a retailer—testifies both to the vast overcapacity of today's private computing systems and to the way utilities can remedy the problem.

Because Amazon is allowing customers to run all sorts of programs on its machines, it can't tailor its system to run a limited number of applications the way Google does. It has to be more flexible. For that reason, Amazon's system works in a way that's very different from Google's. Instead of being based on parallel processing, it's built on a technology known as virtualization—a technology that will be crucial to the future development of utility computing. Indeed, without virtualization, large-scale utility computing is unthinkable.

As with many multisyllabic computer terms, "virtualization" is not quite as complicated as it sounds. It refers to the use of software to simulate hardware. As a simple example, think of the way the telephone answering machine has changed over the years. It began as a bulky, stand-alone box that recorded voices as analog signals on spools of tape. But as computer chips advanced, the answering machine turned into a tiny digital box, often incorporated into a phone. Messages weren't inscribed on tape, but rather stored as strings of binary bits in the device's memory. Once the machine had become fully digitized, though, it no longer had to be a machine at all. All its functions could be replicated through software code. And that's exactly what happened. The box disappeared. The physical machine turned into a virtual machine—into pure software running out somewhere on a phone company's network. Once you had to buy an answering machine. Now you can subscribe to an answering service. That's the essence of virtualization.

Because all the components of computer systems, from microprocessors to storage drives to networking gear like routers, firewalls, and load balancers, operate digitally, they too can be replaced by software. They can be virtualized. When you rent a computer—or a thousand of them, for that matter—through Amazon's EC2 service, you're not renting real computers. You're renting virtual machines that exist only in the memory of Amazon's physical computers. Through virtualization, a single Amazon computer can be programmed to act as if it were many different computers, and each of them can be controlled by a different customer.

Virtualization has long been an important part of computing. It was one of the technologies that allowed mainframes to handle a lot of different jobs simultaneously. But today it has become truly revolutionary, thanks to the explosion in the power of computer chips. Because running a virtual machine is no different from running a software application, it consumes a good deal of a microprocessor's power. Until recently, that limited virtualization's usefulness. Running just one or two virtual machines would slow a computer to a crawl—there'd be no processing capacity left to do anything with the machines. But ordinary microprocessors have become so powerful that they can run many virtual machines simultaneously while still having plenty of power in reserve to run sophisticated business applications on each of those machines.

Virtualization breaks the lock between software and hardware that made client–server computing so inefficient and complicated. A company no longer has to dedicate a powerful server to just one application. It can run many applications with the computer, and it can even automatically shift the machine's capacity between the applications as demand for them fluctuates. Virtualization allows companies—or the utilities that serve them—to regain the high capacity utilization that characterized the mainframe age while gaining even more flexibility than they had during the PC age. It offers the best of both worlds.

Virtualized systems that are shared by many companies are often referred to by computer professionals as "multi-tenant systems." The name suggests a metaphor that gets at an essential difference between the client–server and the utility models of computing. When you install a new system in the client–server model, you have to build the equivalent of, say, a four-story building—but the building ends up being occupied by just a single tenant. Most of the space is wasted. With the use of virtualization in the utility model, that building can be divided into apartments that can be rented out to dozens of tenants. Each tenant can do whatever it likes inside the walls of its own apartment, but all of them share the building's physical infrastructure—and all of them enjoy the resulting savings.

The idea of multi-tenancy also reveals why utility computing is fundamentally different from systems outsourcing, the traditional way companies have offloaded some of the burden of maintaining their own information technology. Both utility computing and outsourcing involve hiring an outside firm to supply computing-related services, but that's where the similarity ends. In an outsourcing arrangement, the supplier simply manages a traditional client–server setup on behalf of its customer. The hardware and software are still dedicated to that one client—and in many cases they're still owned by the client as well. The outsourcer may achieve some savings in labor costs, but the underlying inefficiency of the client–server model remains.

Another company that's using virtualization as the basis for multi-tenant services is Savvis Inc. Founded in St. Louis in 1995, Savvis originally had two major businesses. It was a large Internet service provider, selling network bandwidth and Internet connections to corporations, and it was also a hosting company that operated big, shared data centers in which companies could house their computers and related gear. But after barely surviving the dotcom shakeout, Savvis came to see that it could use virtualization to combine those two services into a full-fledged utility.

Unlike Google and Amazon, which built their utility systems out of cheap commodity equipment, Savvis chose to buy expensive, cutting-edge gear like Egenera servers and 3PAR storage systems. Egenera and 3PAR are small companies that specialize in building highly reliable hardware designed specifically for the virtual systems run by large-scale utilities. The added cost of the equipment makes the strategy risky, but it has enabled Savvis to automate the deployment of information technology to a degree that hasn't been possible before. A company using Savvis's utility doesn't have to worry about setting up individual virtual machines. It just tells Savvis what its maximum computing requirement is, and the system creates virtual machines and transfers applications between them in response to shifts in demand. Each client's usage of the system is tracked automatically and documented in a monthly bill.

The ultimate goal, according to Savvis's chief technology officer, Bryan Doerr, is not just to virtualize computers and other components but to create an entire virtual data center, encompassing computing, storage, and networking. The center could be encapsulated and stored, literally, in a single digital file. You could thus launch a new center as easily as you launch a software application today. Managing an entire corporate computing operation would require just one person sitting at a PC and issuing simple commands over the Internet to a distant utility.

That may sound outlandish to anyone who has struggled to put together the many pieces of software and hardware required to run a business application. But it's rapidly becoming a reality. Late in 2006, an innovative California startup called 3Tera introduced a software program, AppLogic, that automates the creation and management of complex corporate systems. Using a simple graphical interface, a system designer can drag and drop icons representing traditional components—servers, databases, routers, firewalls, and the cables that connect them—onto a page in a Web browser. When he has the arrangement he wants, he clicks a button, and AppLogic builds the system, virtually, on a grid of generic computers at a utility. What once required big outlays of cash and days of work can now be done in minutes without buying any hardware or hiring any technicians. Built into AppLogic is a meter that measures usage. A customer pays only for the computing power it consumes, when it consumes it.

3Tera's software provides a hint of what the future of the computer business may look like. The makers of many kinds of gear may not have to manufacture physical products at all. They may create virtual versions of their equipment entirely out of software and sell them as icons that can be plugged into programs like AppLogic. Of course, that raises an even more radical possibility. Instead of being made and sold by hardware vendors, the virtual devices could be incorporated into applications or even produced as open-source software and given away free. Much of the traditional hardware business would simply disappear.

IN THE MOST radical version of utility computing available today, the utility displaces the personal computer entirely. Everything that a person does with a PC, from storing files to running applications, is supplied over the computing grid instead. Rendered obsolete, the traditional PC is replaced by a simple terminal—a "thin client" that's little more than a monitor hooked up to the Internet. Thin clients have been around for years, and they've become increasingly popular in the corporate market, where their recent sales growth, at more than 20 percent a year, far outpaces that of PCs. Companies have found that the stripped-down machines are ideal for employees with narrowly focused jobs, such as customer service representatives, reservation agents, and bank tellers. Because such workers generally require only a few software programs, they have little need for a multipurpose PC. By supplying the applications and data over a network, companies can avoid most of the maintenance and other costs associated with traditional PCs and their complicated software.

The thin-client model has particular appeal in the developing world, where millions of businesses, schools, and individuals can't afford to purchase even the lowest-priced PCs. In India, for example, a company named Novatium is having success providing personal computing as a simple utility service. Its customers receive a thin client, called a Nova netPC, as well as a set of software services, all provided through their local telephone companies and paid for through a small charge on their phone bills. Household accounts also receive an hour of free Internet access a day. School and business accounts have various additional software and Internet options to choose from, at different prices. Not only do customers avoid the cost of buying a PC, but they also avoid all the hassles that go along with PC ownership, from installing and upgrading software to troubleshooting problems to fighting viruses.

Today, it's hard to imagine computer owners in the United States and other developed countries abandoning their PCs for thin clients. Many of us, after all, have dozens or even hundreds of gigabytes of data on our personal hard drives, including hefty music and video files. But once utility services mature, the idea of getting rid of your PC will become much more attractive. At that point, each of us will have access to virtually unlimited online storage as well as a rich array of software services. We'll also be tapping into the Net through many different devices, from mobile phones to televisions, and we'll want to have all of them share our data and applications. Having our files and software locked into our PC's hard drive will be an unnecessary nuisance. Companies like Google and Yahoo will likely be eager to supply us with all-purpose utility services, possibly including thin-client devices, for free—in return for the privilege of showing us advertisements. We may find, twenty or so years from now, that the personal computer has become a museum piece, a reminder of a curious time when all of us were forced to be amateur computer technicians.

THE MEMO Bill Gates sent out in late October 2005 was actually just a cover note. Attached was a much longer document that laid out Microsoft's plan for making the transition from the PC age to the utility age. It was written not by Gates but by Ray Ozzie, a distinguished software engineer who had joined Microsoft a year earlier and taken over Gates's post as chief designer of the firm's software. Ozzie was even more emphatic than his boss about the revolutionary potential of what he termed the "Internet Services Disruption." "The environment has changed yet again," he wrote. "Computing and communications technologies have dramatically and progressively improved to enable the viability of a services-based model. The ubiquity of broadband and wireless networking has changed the nature of how people interact, and they're increasingly drawn toward the simplicity of services and service-enabled software that 'just works.'" Microsoft's mainstream business customers, he wrote, "are increasingly considering what services-based economies of scale might do to help them reduce infrastructure costs or deploy solutions as-needed and on [a] subscription basis." It was imperative, he concluded, "for each of us to internalize the transformative and disruptive potential of services."

Ozzie's memo, like Gates's, made it clear that Microsoft had no intention of surrendering. It would turn its legendary competitiveness against the likes of Google and Salesforce.com in hopes of sustaining its dominance through the transition to the utility age. The company soon let it be known that it would launch an aggressive program of capital investment to expand its utility computing capabilities in an effort to catch up with Google. The amounts would be "staggering," Ozzie told a reporter from *Fortune*. In 2006 alone, Microsoft invested $2 billion more than it had previously expected to spend, much of it going to build and outfit new utility data centers. The most ambitious of those centers, encompassing six buildings and 1.5 million square feet of space, was erected near the Columbia River in the potato-farming town of Quincy, Washington—just a hundred miles upstream from Google's The Dalles plant. Microsoft is using its new data centers to deliver an array of utility services to businesses and consumers, through its Windows Live, Office Live, and MSN brands. But pioneering a new business while continuing to harvest profits from an old one is one of the toughest challenges a company can face. It remains to be seen whether Microsoft will be able to pull it off.

Other big tech companies also recognize that their businesses are under threat, and they too are taking steps to adapt to the utility age. Oracle has begun offering Web-based software services in addition to its traditional applications. SAP has a partnership with Deutsche Telekom that allows clients to run its applications on a utility grid operated by the German telecommunications giant's T-Services unit. IBM and Hewlett–Packard have set up utility data centers to provide computing power for a metered fee, and HP has acquired a large thin-client manufacturer named Neoware. Sun has developed a new generation of energy-efficient computers geared to the needs of utility operators. EMC, a vendor of traditional storage systems, owns a controlling stake in VMware, the leading supplier of virtualization software. Even large IT consulting firms like Accenture, which grew rich on the complexity of traditional information systems, are establishing practices devoted to helping clients shift to utility services. Nobody wants to be left in the dust.

Some of the old-line companies will succeed in making the switch to the new model of computing; others will fail. But all of them would be wise to study the examples of General Electric and Westinghouse. A hundred years ago, both these companies were making a lot of money selling electricity-production components and systems to individual companies. That business disappeared as big utilities took over electricity supply. But GE and Westinghouse were able to reinvent themselves. They became leading suppliers of generators and other equipment to the new utilities, and they also operated or invested in utilities themselves. Most important of all, they built vast new businesses supplying electric appliances to consumers—businesses that only became possible after the arrival of large-scale electric utilities. Sometimes a company can discover an even better business if it's willing to abandon an old one.

On June 15, 2006, Microsoft announced in a press release that Bill Gates would be stepping down from his managerial role. He would hand off his remaining responsibilities to Ozzie and other executives, and then, in 2008, he would retire from his day-to-day work at the company that he had built into the superpower of the PC age. Gates's departure is richly symbolic. It marks, more clearly than any other event, a turning point in the brief but tumultuous history of computing. The time of Gates and the other great software programmers who wrote the code of the PC age has come to an end. The future of computing belongs to the new utilitarians.

CHAPTER 5

The White City

N 1893, JUST a year after Samuel Insull arrived in Chicago, the city hosted the largest world's fair ever—the Columbian Exposition—to commemorate the four hundredth anniversary of Christopher Columbus's voyage to the New World. Constructed on a 633-acre site on the shores of Lake Michigan, the fairgrounds formed a spectacular city-within-a-city, featuring ornate neoclassical exhibition halls, displays by dozens of countries and hundreds of businesses, and a midway with a 265-foot-tall Ferris wheel, the first ever built.* The fair drew more than 27 million visitors—equal to nearly half of the country's population at the time—over the course of its five-month run.

The Columbian Exposition was a monument to the idea of technological progress. It celebrated advances in industry, transportation, and the arts, but most of all it celebrated the arrival of electricity as the nation's new animating force. The organizers of the event wrote that it was their intention "to make the World's Fair site and the

* It is believed that the model for George W. G. Ferris's beloved invention was Henry Burden's waterwheel. Ferris was educated at the Rensselaer Polytechnic Institute in Troy, just a few miles from Burden's ironworks.

buildings one grand exemplification of the progress that has been made in electricity." A steam plant, built on the grounds, pumped out 24,000 horsepower of energy, nearly three-quarters of which went to generating electric current. During its run, the exposition consumed three times as much electricity as the rest of the city of Chicago.

The electricity powered railroads and boats, moving walkways, elaborate fountains, and scores of commercial displays of the latest machinery and appliances. Most of the current, though, went into the 100,000 incandescent bulbs, arc lights, and neon tubes that illuminated the "White City," as the fairgrounds came to be called. One visitor described the sight of the exposition at night in rhapsodic terms: "The gleaming lights outlined the porticoes and roofs in constellations, studded the lofty domes with fiery, falling drops, pinned the darkened sky to the fairy white city, and fastened the city's base to the black lagoon with nails of gold." The beam of a searchlight, he wrote, "flowed toward heaven until it seemed the holy light from the rapt gaze of a saint or Faith's white, pointing finger." Another visitor, L. Frank Baum, was so dazzled by the fair that it became the inspiration for the Emerald City in his 1900 book *The Wonderful Wizard of Oz*.

One of the exposition's most popular attractions was the palatial Electricity Building. Sprawling over five and a half acres, it contained, in addition to an 80-foot-tall Tower of Light built by Thomas Edison, hundreds of exhibits of the latest electrical equipment, including "horseless carriages" powered by batteries. An awestruck Henry Adams spent two weeks exploring the treasures of the Columbian Exposition, but he was most deeply affected by seeing a display of electric dynamos—two 800-kilowatt General Electric machines, the largest available at the time, and a dozen of the latest Westinghouse generators. He recalled the experience in his autobiography *The Education of Henry Adams*. "One lingered long among the dynamos," he wrote, "for they were new, and they gave to history a new phase." Sensing that such machines "would result in infinite costless energy within a generation," Adams knew that they would reshape the country and the world. He felt humbled by the dynamos, but their power also troubled him. What history's "new phase" would bring, he realized, lay beyond our understanding and even our control: "Chicago asked in 1893 for the first time the question whether the American people knew where they were driving."

SAMUEL INSULL WAS a visionary, but not even he could have imagined how profoundly, and quickly, the electric grid would reshape business and society. The essence of the new technology's transformative power lay in the way it changed the economic trade-offs which influence, often without our awareness, the many small and large decisions we make that together determine who we are and what we do—decisions about education, housing, work, family, entertainment, and so on. In short, the central supply of cheap electricity altered the economics of everyday life. What had been scarce—the energy needed to power industrial machines, run household appliances, light lights—became abundant. It was as if a great dam had given way, releasing, at long last, the full force of the Industrial Revolution.

If no visionary could have foreseen the course and extent of the changes in store, that didn't prevent many would-be prophets from trying. In the wake of the Chicago fair, electricity took hold of the public imagination. At once glamorous and dangerous, it seemed, in itself and even more so in its applications, like a mysterious, invisible force that had leapt into the world out of the pages of science fiction. Writers and lecturers vied to sketch out the most wondrous scenario of an electrified future. Some of the scenarios were dark, but most were optimistic, often extravagantly so. The last few years of the nineteenth century saw the publication of more than 150 books predicting the imminent arrival of a technological paradise, and utopian literature remained popular throughout the early decades of the twentieth century, when the wires of the electric grid were being strung across the country. Electricity production managed to bring into harmony, at least briefly, two discordant themes running through American culture: utilitarianism and transcendentalism.

Electrification, people were told, would cleanse the earth of disease and strife, turning it into a pristine new Eden. "We are soon to have everywhere," wrote one futurist, "smoke annihilators, dust absorbers, ozonators, sterilizers of water, air, food, and clothing, and accident preventers on streets, elevated roads, and subways. It will become next to impossible to contract disease germs or get hurt in the city." Another announced that "electrified water" would become "the most powerful of disinfectants." Sprayed into "every crack and crevice," it would obliterate "the very germs of unclean matter." "Indeed," wrote another, "by the all potent power of electricity, man is now able to convert an entire continent into a tropical garden at his pleasure."

Electrified machines would eliminate blizzards, droughts, and other climatic extremes, giving man "absolute control of the weather." Inside homes, "electric equalizers" would send out a "soothing magnetic current" to "dissipate any domestic storm and ensure harmony in families." New transportation and communication systems would "practically eliminate distances," just as electric lights would abolish "the alternation of day and night." Eventually, the "human machine" would be "thoroughly understood and developed to its highest efficiency." And then all the individual human machines would join together to form an even greater machine.

People would become "cogs" in a "wonderful mechanism . . . acting in response to the will of a corporate mind as fingers move and write at the direction of the brain."

The utopian rhetoric was not just a literary conceit. It proved a powerful marketing pitch for the manufacturers of electric appliances. General Electric was particularly adept at playing to people's native optimism about technology. During the 1920s, the decade in which the pace of wiring American homes reached its peak, the company increased its yearly promotional expenditures from $2 million to $12 million. It devoted much of the money to instilling in the public mind what it called "a positive electrical consciousness" through a concerted program of magazine advertisements, pamphlets, and presentations at schools and women's clubs. Typical of the campaign was a booklet called "The Home of a Hundred Comforts," which described with flowery prose and futuristic illustrations how electric appliances would eliminate most household work, bestowing a life of ease and leisure on formerly harried homemakers. Having electricity in a house, the company's marketers proclaimed, would be like having "ten home servants."

Whether conjured up for literary or commercial purposes, the utopian future never arrived. Cheap electricity brought great benefits to many people, but its effects rarely played out as expected, and not all of them were salubrious. Tracing the course of some of the most important of those effects through the first half of the last century reveals the complex interplay between technological and economic systems and the equally complex way it exerts its influence over society.

BUSINESSES FELT electrification's most immediate impact. With electric light and power, factory owners could build bigger and more productive plants, boosting their output and gaining greater advantages of scale over smaller businesses. In many industries manufacturers rushed to merge with other manufacturers, consolidating production capacity into the hands of an ever-shrinking number of large companies. These big businesses coordinated and controlled their far-flung operations with new communication and computation technologies that also relied on electricity, including the telephone network and the punch-card tabulator. The modern corporation, with its elaborate management bureaucracy, emerged in its familiar form and rose to its dominant position in the economy.

But while electrification propelled some industries to rapid growth, it wiped out others entirely. During the 1800s, American companies had turned the distribution of ice into a thriving worldwide business. Huge sheets were sawn from lakes and rivers in northern states during the winter and stored in insulated icehouses. Packed in hay and tree bark, the ice was shipped in railcars or the holds of schooners to customers as far away as India and Singapore, who used it to chill drinks, preserve food, and make ice cream. At the trade's peak, around 1880, America's many "frozen water companies" were harvesting some 10 million tons of ice a year and earning millions in profits. Along Maine's Kennebec River alone, thirty-six companies operated fifty-three icehouses with a total capacity of a million tons. But over the next few decades, cheap electricity devastated the business, first by making the artificial production of ice more economical and then by spurring homeowners to replace their iceboxes with electric refrigerators. As Gavin Weightman writes in *The Frozen-Water Trade,* the "huge industry simply melted away."

Inside companies, things changed as well. As manufacturers replaced their millwork and gas lamps with electric motors and lightbulbs, working conditions improved substantially. Gone were the elaborate gearing systems that leaked grease and oil onto workers and their machines. Gone were the foul and often sickening fumes let off by gas flames. The stronger and steadier illumination provided by incandescent lights reduced accidents and eased the strain of close work. Electric fans brought in fresh air. Although electric machines and automated factories would bring new dangers, like the risk of electrocution, the health and productivity of factory workers in general improved.

But as working conditions got better, work itself underwent a less benign transformation. For two centuries, since the invention of the steam engine launched the Industrial Revolution, mechanization had been steadily reducing the demand for talented craftsmen. Their work had been taken over by machines that required little skill or training to operate. Electricity accelerated the trend. Because electric current could be regulated far more precisely than power supplied through shafts and gears, it became possible to introduce a much wider range of industrial machines, leading to a further "deskilling" of the workplace. Factory output skyrocketed, but jobs became mindless, repetitious, and dull. In many cases, the very movements of workers came to be carefully scripted by industrial engineers like Frederick W. Taylor, who used stopwatches and motion studies to ensure that work proceeded with scientific efficiency. Industrial workers did indeed become "cogs" controlled by "the will of a corporate mind." The experience, though, was anything but heavenly.

Mass production found its fulfillment in the creation of the modern assembly line, an innovation that would have been unthinkable before electrification. The automated line made its debut in 1913 in Henry Ford's Highland Park plant, which, as the historian David Nye describes in his book *Consuming Power,* had been constructed "on the assumption that electrical power should be available everywhere." Electricity and electric motors provided Ford and other manufacturers with far more control over the specification, sequencing, and pacing of tasks. They made it possible to build highly specialized machines, arrange them in the best possible order, and connect them with a conveyor belt that could run at variable speeds. At the same time, electric machine tools made it possible to mass-produce individual parts to a set of rigid specifications. Interchangeable parts were essential to the operation of assembly lines.

Ford himself would later stress the critical role that electric utilities played in paving the way for his revolutionary plant:

> The provision of a whole new system of electric generation emancipated industry from the leather belt and line shaft, for it eventually became possible to provide each tool with its own electric motor. . . . The motor enabled machinery to be arranged according to the sequence of work, and that alone has probably doubled the efficiency of industry, for it has cut out a tremendous amount of useless handling and hauling. The belt and line shaft were also very wasteful of power—so wasteful, indeed, that no factory could be really large, for even the longest line shaft was small according to modern requirements. Also high-speed tools were impossible under the old conditions—neither the pulleys nor the belts could stand modern speeds. Without high-speed tools and the finer steels which they brought about, there could be nothing of what we call modern industry.

Widely adopted by other manufacturers, Ford's electrified assembly line brought a big leap in industrial productivity. In 1912, it took 1,260 man-hours to produce a Model T. Two years later, with the assembly line in operation, that number had been halved, to 617 man-hours. As the line's operation continued to be fine-tuned, the labor requirement dropped further, falling to just 228 man-hours by 1923. By streamlining and speeding up the production process, the assembly line also sharply reduced the inventories of components and half-built products that had to be kept on factory floors. Manufacturing became far more profitable. While much of the new profits went to companies' owners or stockholders, a substantial portion ended up in workers' pockets.

Just as he had pioneered the assembly line, Ford also led the way in boosting blue-collar wages. Soon after opening the Highland Park plant, he announced he would double workers' pay—to five dollars a day—across the board. Although the move earned him the ire of other businessmen and a dressing-down from the *Wall Street Journal*, Ford saw that higher wages were necessary to convince large numbers of men to take factory jobs that had become numbingly tedious—and to discourage them from quitting those jobs after a brief tenure. In response to the pay hike, 15,000 men lined up outside Ford's factory to apply for 3,000 open slots. Other factory owners soon realized they had little choice but to keep pace with Ford, and they too began raising wages. The rapid increase in the sizes of plants and their workforces also accelerated the spread of unionization—a development Ford fought, sometimes brutally—which helped funnel even more profit from the owners to the hands.

Here was the first, but by no means the last, irony of electrification: even as factory jobs came to require less skill, they began to pay higher wages. And that helped set in motion one of the most important social developments of the century: the creation of a vast, prosperous American middle class.

Another development in the labor market also played an important role in the rise of the middle class. As companies expanded, adopted more complicated processes, and sold their goods to larger markets, they had to recruit more managers and supervisors to oversee and coordinate their work. And they had to bring in many other kinds of white-collar workers to keep their books, sell their goods, create marketing and advertising campaigns, design new products, recruit and pay employees, negotiate contracts, type and file documents, and, of course, operate punch-card tabulators and related business machines. As industries such as chemicals manufacturing and steel-making became more technologically advanced, moreover, companies had to hire cadres of scientists and engineers. While the expansion of the white-collar workforce, like the mechanization of factories, began before electrification, cheap power accelerated the trend. And all the new office jobs paid well, at least by historical standards.

The shift in skilled employment away from tradesmen and toward what would come to be known as "knowledge workers" had a knock-on effect that also proved pivotal in reshaping American society: it increased the workforce's educational requirements. Learning the three Rs in grammar school was no longer enough. Children needed further and more specialized education to prepare them for the new white-collar jobs. That led to what Harvard economist Claudia Goldin has termed "the great transformation of American education," in which public education was extended from elementary schools to high schools. Secondary education had been rare up through the early years of the century; it was reserved for a tiny elite as a preparatory step before entering university. In 1910, high-school enrollment in even the wealthiest and most industrially advanced regions of the country rarely exceeded 30 percent of 14- to 17-year-olds, and it was often considerably lower than that. But just twenty-five years later, average enrollment rates had jumped to between 70 and 90 percent in most parts of the country. Going to high school, which a generation earlier wouldn't have entered the minds of most kids, had become a routine stop on the way to a decent job.

IF ELECTRIFICATION HELPED spur the development of a large and increasingly well-educated middle class, the new middle class in turn helped extend electrification's reach and amplify its impact. Both blue- and white-collar workers spent their wages on the new electric appliances being produced by the electrified manufacturing companies that employed them. The burgeoning demand helped the factories achieve even greater economies of scale, pushing down the prices of the products and generating even more sales. This economic cycle ended up creating huge markets for all sorts of appliances, which in turn reinforced the virtuous cycle that electric utilities had enjoyed since factories began buying their power: more appliances meant more consumption of electricity, which led to even greater economies for the utilities, allowing them to cut electric rates still further and spur even more demand for their current and the appliances that ran on it.

Critical to this process was the standardization of the electric system. Although electric utilities originally produced current in a wide variety of frequencies and voltages, both the utilities and the new appliance manufacturers quickly came to see the benefits of universal standards—not just for the current but also for motors, wires, transformers, sockets, and other components of the system. Without standards, without a grid that truly acted as one machine, motors and appliances would have to be designed differently for different markets, reducing scale economies in production, keeping prices high, and curtailing sales. Weak demand for appliances would, in turn, dampen demand for current. To avoid such a scenario, utilities and manufacturers established associations—the National Electric Light Association and the National Electrical Manufacturers Association, respectively—to hash out standards and promote, with the government's encouragement, the sharing of patents through cross-licensing. Soon, electricity was being generated everywhere at 60 cycles and delivered through wall sockets in homes and offices at a steady 120 volts.

The two interlocking economic cycles—one spurring demand for electric power, the other for electric appliances—set off a series of changes in Americans' domestic and leisure lives that were every bit as dramatic as those unfolding in their work lives. The first electric appliance to be bought in large quantities, other than the incandescent lamp, was the electric fan. A simple machine, it made life a little more bearable during summer heat waves. But as new and more sophisticated electric appliances proliferated in the early years of the new century, they began to alter people's behavior and expectations. They opened up to the masses activities that had once been limited to the few, and they made entirely new kinds of social and cultural experiences possible. Cities lit up, literally and figuratively, with dazzling new attractions. "Here is our poetry," wrote an awestruck Ezra Pound when, in 1910, he gazed for the first time upon Manhattan's nighttime illuminations, "for we have pulled down the stars to our will."

The profusion of possibilities for spending one's time and money changed people's very conception of leisure and entertainment, creating a new kind of popular culture and making consumption and consumerism egalitarian pursuits. David Nye's bird's-eye survey of the myriad cultural by-products of electrification captures the scope of the phenomenon well:

Electricity made possible the radio, the telephone, motion pictures, the microphone, the amplifier, the loudspeaker, the trolley, and the spectacular lighting displays in theaters and along the Great White Way [of the urban main street]. Less obvious but no less important, electricity made available artificial daylight, precise delivery of heat and ventilation, the escalator, and the elevator. Americans used it to create new urban environments: the skyscraper, the department store, the amusement park . . . The sober conservation of energy no longer seemed necessary in a world where the power supply seemed unlimited.

Electrification's immediate effects did not always turn out to be its lasting effects, however. While inexpensive power initially promoted mass entertainment on a scale never seen before, drawing throngs of people into the bright and bustling centers of cities, its long-term impact was very different, due in large measure to the emergence of the automobile as the primary means of transport. Although cars weren't electrical products themselves, it was the electrified assembly line that made them cheap enough to be purchased by average citizens. Before Ford's Highland Park plant, cars were built painstakingly by hand, each part custom-fashioned by craftsmen. They were exotic luxury goods affordable only to the wealthy. But Ford's plant could produce standardized Model Ts so inexpensively and in such quantity that, by the early 1920s, dealers were selling millions of them for just $290 apiece. Americans abandoned electric trolleys and trams and took to the road in their own private cars. That spurred heavy investments in oil fields, refineries, and filling stations, flooding the market with inexpensive gasoline and making driving even more attractive.

Cheap electricity, cheap cars, and cheap gas, combined with the rising incomes of a growing middle class, prompted an exodus from cities to suburbs and a shift from the public entertainments offered by theaters, amusement parks, and urban streets to the private diversions served up by televisions, radios, and hi-fi sets. Broadcast media, also made possible by electricity, brought the Great White Way of the city into the living room—and, thanks to advertising, you didn't even have to buy a ticket. The spectacle came to you, conveniently and for free. The mass culture remained, and in fact was further strengthened by popular radio and television programs and hit records, but its products were increasingly consumed in private.

THE HOME WASN'T just a setting for recreation and relaxation. It was also a place of work, work done mainly by women. In the early decades of the twentieth century, few middle-class women held jobs outside the home. In the city of Muncie, Indiana, for example, a 1925 survey found that fewer than three in a hundred women worked for pay; the rest were homemakers. Before electrification, many common household chores were performed in uncomfortable conditions and demanded considerable strength and stamina. Even women in households of modest means would hire servants or pay washerwomen or other day laborers to shoulder some of the burden. The men of the house would also pitch in on the heavier jobs, such as dragging rugs outside to be beaten or hauling and heating tubs of water for laundry.

The utopian promise of electricity seemed within reach inside the home. Many women believed that new appliances like vacuum cleaners and washing machines would, as General Electric advertised, transform their houses from places of labor into places of ease. The home would become less like a sweatshop and more like a modern, automated factory, and the housewife would become, as Thomas Edison predicted in a 1912 article on "The Future of Women," "a domestic engineer [rather] than a domestic laborer, with the greatest of handmaidens, electricity, at her service." The first widely purchased appliance designed specifically for housework, the electric iron, seemed to fulfill this expectation. Women no longer had to heat a heavy wedge of cast iron over a hot stove and then drag the red-hot chunk of metal over a piece of clothing, stopping frequently to reheat it. They could just plug a lightweight appliance into the wall. During the first two decades of the century, scores of homemakers swapped their old-fashioned irons for modern electric ones. A photograph of the time shows a General Electric employee standing proudly beside a small mountain of discarded flat irons.

As it turned out, though, the electric iron was not quite the unalloyed blessing it first appeared to be. By making ironing "easier," the new appliance ended up producing a change in the prevailing social expectations about clothing. To appear respectable, men's and women's blouses and trousers had to be more frequently and meticulously pressed than was considered necessary before. Wrinkles became a sign of sloth. Even children's school clothes were expected to be neatly ironed. While women didn't have to work as hard to do their ironing, they had to do more of it, more often, and with more precision.

As other electric appliances flooded the home through the first half of the century—washing machines, vacuum cleaners, sewing machines, toasters, coffee-makers, egg beaters, hair curlers, and, somewhat later, refrigerators, dishwashers, and clothes dryers—similar changes in social norms played out. Clothes had to be changed more frequently, rugs had to be cleaner, curls in hair had to be bouncier, meals had to be more elaborate, and the household china had to be more plentiful and gleam more brightly. Tasks that once had been done every few months now had to be performed every few days. When rugs had had to be carried outside to be cleaned, for instance, the job was done only a couple of times of year. With a vacuum cleaner handy, it became a weekly or even a daily ritual.

At the same time, the easing of the physical demands of housework meant that many women no longer felt justified in keeping servants or hiring day workers. They felt able—and thus obligated—to do everything themselves. (And many of the poor women who had been servants were moving into higher-paying factory jobs, anyway.) Housewives also lost the helping hands that had been provided by their husbands and sons, who, now that the work was "easy," no longer felt obliged to pitch in. In middle-class households, reports Ruth Schwartz Cowan in *More Work for Mother,* "the labor saved by labor-saving devices was that not of the housewife but of her helpers."

A series of studies of the time women devoted to housework back up Cowan's observation. Research undertaken between 1912 and 1914, before the widespread adoption of electric appliances, found that the average woman spent 56 hours a week on housework. Similar studies undertaken in 1925 and 1931, after electric appliances had become common, found that they were still spending between 50 and 60 hours a week on domestic chores. A 1965 study again found little change—women were spending on average 54.5 hours per week on housework. A more recent study, published in 2006 by the National Bureau of Economic Research, also found that the hours housewives devoted to domestic work remained steady, at between 51 and 56 a week, in every decade from the 1910s through the 1960s.

Electrification changed the nature of "women's work," in other words, but it didn't reduce its quantity. One of the pamphlets General Electric produced to promote its appliances was titled "A Man's Castle Is a Woman's Factory." But in that factory, women didn't just assume the role of the manager or engineer. They also took on, usually alone, the role of the unskilled machine operator. Far from delivering a life of leisure to middle-class homemakers, electrification actually resulted in what Cowan terms the "proletarianization" of their work.

It didn't necessarily feel like that, though, to the women loading the washing machines and pushing the vacuum cleaners. In their advertising and public relations campaigns, utilities and appliance manufacturers promoted the adoption of electric appliances as a civilizing force in society. It was a pitch that both reflected and reinforced a deep change in attitudes about the role of the homemaker. The idea of "home economics" became popular at this time, with its aim of, as one contemporary speaker put it, "bring[ing] the home into harmony with industrial conditions and social ideas." To buy and use electric appliances—to be a dedicated consumer of electricity—was to further progress, to help bring about a more modern and a better world. The housewife, like the factory hand, had become an essential cog in the great technological machine that was producing a more advanced civilization.

Before the arrival of electric appliances, homemaking had been viewed as work—as a series of largely unpleasant but inescapable tasks. If it wasn't always drudgery, it was always something that *had* to be done, not something that one would have chosen to do. After electrification, homemaking took on a very different character. It came to be seen not as a chore but as a source of identity and, in itself, a means of personal fulfillment. Women saw their status and their worth as being inextricably linked to their success as a homemaker, which in turn hinged on their ability to master domestic machinery. The new feminine ideal was promulgated not only by ads but by articles and photographs in popular magazines like *Ladies' Home Journal* and *Good Housekeeping*. As Nye puts it, "The adoption of new technologies in the home was in part a matter of personal prestige and conspicuous consumption, in part an expression of the belief in scientific progress concretized in a new vacuum cleaner or electric refrigerator, and partly an affirmation of new roles for women as home managers." Appliances turned into "tools of psychological maintenance and symbols of transformation."

The psychic price of the new tools and the new roles they engendered was sometimes high, however. Women labored under escalating pressures: to meet the higher expectations for cleanliness and order, to purchase the latest "must-have" appliances, to learn how to operate all the new machines and keep them in working order. And, for many, electrification brought a new sense of alienation and loneliness into the home. As women took over all the work required to keep house, they often found themselves spending more of their time alone, isolated in their suburban residences. They may have had their young children to keep them company, but adult companionship was usually rarer than it had been in the past when homemaking was more of a communal activity. Life had changed, in subtle as well as obvious ways.

IN 1899, A GROUP of Texans founded a professional society for men involved in the new electricity business. Called the Jovians, the group's motto was "All together, all the time, for everything electrical." The fraternity expanded rapidly, attracting thousands of members across the country, from utility executives to linesmen. Thomas Edison, George Westinghouse, and Samuel Insull all became members. One of the society's founders, Elbert Hubbard, wrote an essay in 1913 in which he described the common bond and purpose shared by Jovians. "Electricity occupies the twilight zone between the world of spirit and the world of matter," he wrote. "Electricians are all proud of their business. They should be. God is the Great Electrician."

The conceit, for all its vainglory, was not so far-fetched. As we look back over the last century, we can see that the businessmen and engineers who invented the electric grid and manufactured a multitude of magical new appliances did wield an almost godlike power. As Henry Adams had predicted, a new world was conjured out of their network of wires. Utility-supplied electricity was by no means the only factor behind the great changes that swept American business and culture in the first half of the twentieth century. But whether it exerted its influence directly or through a complicated chain of economic and behavioral reactions, the electric grid was the essential, formative technology of the time—the prime mover that set the great transformations in motion. It's impossible to conceive of modern society taking its current shape—what we now sense to be its natural shape—without the cheap power generated in seemingly unlimited quantities by giant utilities and delivered through a universal network into nearly every factory, office, shop, home, and school in the land. Our society was forged—we were forged—in Samuel Insull's dynamo.

Part 2

Living in

the Cloud

We shape our tools
and thereafter they shape us.
—*John M. Culkin*

World Wide Computer

I F YOU HAD hung out on Haight Street in San Francisco during 1967's Summer of Love, there's a good chance you would have come across the countercultural bard Richard Brautigan reciting an ecstatic ode to a computerized future. The 25-line poem, likely inspired by Brautigan's brief stint earlier that year as poet in residence at the California Institute of Technology, described a "cybernetic meadow" where people and computers would "live together in mutually / programming harmony." It closed with a vision of an entire "cybernetic ecology":

where we are free of our labors
and joined back to nature . . .
and all watched over
by machines of loving grace.

That same year, a group of mathematicians and computer scientists associated with ARPA, the Advanced Research Projects Agency within the US Department of Defense, met at the University of Michigan to begin planning a data communication network that would, as an agency report prosaically put it, "be used for exchanging messages between any pair of computers." The project was aimed at allowing academic, military, and industrial researchers to make more efficient use of scarce and expensive data-processing machines. It was to have two thrusts: the construction of "a 'subnetwork' of telephone circuits and switching nodes whose reliability, delay characteristics, capacity, and cost would facilitate resource sharing among computers on the network," and the design of "protocols and procedures within the operating systems of each connected computer, in order to allow the use of the new subnetwork by the computers in sharing resources." The meeting in Michigan laid the groundwork for the Arpanet, the government-sponsored computer network that would eventually expand into the modern Internet and, finally, into the utility-computing grid. Brautigan's benevolent "cybernetic ecology" turned out to be a creation of the same military-industrial complex that had served as the counterculture's nemesis.

That irony was never allowed to dampen the enthusiasm of Brautigan's heirs—the techno-utopian dreamers who have shaped the public perception of computer systems from the late 1960s to the present day. As computers came to be connected into a single system, many writers and thinkers embraced the view that a more perfect world was coming into being. By linking computers, they saw, you would also link people, creating electronic communities unconstrained by existing social and political boundaries. A technological paradise beckoned. As Fred Turner describes in his book *From Counterculture to Cyberculture,* the most avid of the dreamers tended to be, like Brautigan, associated with the 1960s counterculture or its 1970s successor, the New Age movement. They saw in the "virtual reality" of networked computers a setting for social and personal transcendence. It was a virgin world where they could build a harmonious communal culture and achieve the higher consciousness they sought.

Stewart Brand, the founding editor of the hippie bible *The Whole Earth Catalog,* sounded the keynote for the new techno-utopianism in a 1972 *Rolling Stone* article. He described how a scruffy band of hackers was subverting the Arpanet's military mission by using the network to exchange jokes and play a game called Spacewar. "Ready or not, computers are coming to the people," Brand announced, calling it "good news, maybe the best since psychedelics." The article was remarkably prescient, as Brand foresaw the coming transformation of computers into personal communication devices (and even the rise of online music trading). In his view, people would come to use their computers to escape the control of society's "planners," to become "Computer Bums" liberated from rules and routines. "When computers become available to everybody," he wrote, we will all become "more empowered as individuals and as co-operators."

Fast-forward two decades, to 1990. An idealistic young software programmer named Tim Berners-Lee, working in an office at CERN, the big European physics laboratory straddling the border between Switzerland and France, is writing the codes that will deliver the Internet to the people. He has a vision of a great, unbounded "web of knowledge"—a World Wide Web—that "brings the workings of society closer to the workings of our minds." As the implications of Berners-Lee's invention begin to be understood, the utopian rhetoric surrounding the "cybernetic ecology" takes on an even more euphoric tone, echoing the extravagant predictions heard in the early days of electrification. In a compendium of essays called *Cyberspace: First Steps,* University of Texas professor Michael Benedikt writes of a vision of a "Heavenly City" emerging from the computer network. Cyberspace "opens up a space for collective restoration, and for peace," writes Nicole Stenger, another contributor. "We will all become angels, and for eternity!" The Grateful Dead lyricist John Perry Barlow proclaims the Web "the most transforming technological event since the capture of fire." It is, he says, "the new home of Mind."

But the World Wide Web turned out to be something very different from what Berners-Lee intended and others yearned for. By creating a universal medium, one able not just to display text but to show images and process transactions, the Web transformed the Internet from an intellectual meeting-house into a commercial enterprise. For a brief moment after Berners-Lee unveiled his invention, the Web was largely empty of commercial activity. At the end of 1993, less than 5 percent of sites were in the .com domain. But as the profit-making potential of the new medium became clear, businesses rushed in and commercial sites quickly came to dominate the network. By the end of 1995, half of all sites bore .com addresses, and by mid-1996 commercial sites represented nearly 70 percent of the total. Three decades after the Summer of Love, young people began flocking to San Francisco once again, but they didn't come to listen to free verse or to drop acid. They came to make a killing. The Web had turned out to be less the new home of Mind than the new home of Business.

The Internet has always been a machine of many contradictions, both in the way it works and in the way it is used and perceived. It's an instrument of bureaucratic control and of personal liberation, a conduit of communal ideals and of corporate profits. These and other technical, economic, and social tensions are becoming even more pronounced as the Net turns into a worldwide computing grid and its uses as a general purpose technology multiply. The resolution of the tensions, for good or ill, will determine how the grid's consequences play out over the coming years and decades.

Those consequences are the subject of the rest of this book. While it may not be possible to plot in advance the twists and turns of our computerized future, the past and the present provide important clues. Utility computing is a new and disruptive force, but it's not without precedent. Even before the first electronic computer was built, data-processing machines had been reducing the cost of mathematical calculations, information storage, and communications. By bringing those costs down much further—to zero, in many cases—the supply of computing as a utility will amplify the benefits, and exacerbate the strains, that have always been associated with the automation of information processing. If the electric utility completed what the steam engine began, the computing utility promises to complete what began with Herman Hollerith's punch-card tabulator.

Up to now, this book has focused on computing's supply side—on the business of delivering information technology to the people and companies that use it. Now, the focus will shift to the demand side—to the users themselves and how their behavior is changing as they adapt to the services supplied over the computing grid and navigate the resulting economic, political, and social upheaval. Before we can understand the implications for users, though, we first need to understand how computing is *not* like electricity, for the differences between the two technologies are as revealing as their similarities.

WITH THE ELECTRIC grid, we've always known precisely where to place the socket. It goes between the point where the current is generated and the point where the current is applied to do something useful. The utilities themselves have just two roles, both clearly delineated: they produce electricity and they distribute it. The means of generating power may be diverse, ranging from giant hydroelectric dams or nuclear stations to smaller coal- or oil-burning plants to even smaller cogeneration facilities or wind farms, but they do essentially the same thing: they convert mechanical energy to electricity. All the applications of the electricity are the responsibility not of the utilities but of the utilities' customers. Because the applications are carried out by physical appliances, they can't be delivered over the network from remote sites. Running a vacuum cleaner in a power plant wouldn't help to clean the rugs in your home. The clear divide between electricity's generating infrastructure and its applications—a divide made manifest in the electrical socket—makes the utility model relatively simple when it comes to electric power. Electricity's complexities lie in its applications, and those applications lie outside the sphere of the utility.

Computing is different. Because its applications are delivered through software, they too can be transmitted over the grid as utility services. Unlike a vacuum cleaner, a software program can be shared by many people, all using it at the same time. It doesn't have to run locally or be dedicated to a single user. Computing applications, in contrast to electrical appliances, can thus benefit from the economies of scale that utilities achieve. The prices of appliances have certainly been driven down by the technologies of mass production, but because they're physical devices there's always been a limit to how cheap they can get. That in turn has constrained the purposes to which electric current can be applied. When applications have no physical form, when they can be delivered as digital services over a network, the constraints disappear.

Computing is also much more modular than electricity generation. Not only can applications be provided by different utilities, but even the basic building blocks of computing—data storage, data processing, data transmission—can be broken up into different services supplied from different locations by different companies. Modularity reduces the likelihood that the new utilities will form service monopolies, and it gives us, as the users of utility computing, a virtually unlimited array of options. We can, for instance, use one service to store our data, another to keep our data secure, another to run our data through a microprocessor, and many others to apply our data to particular chores or tasks. The public comput-ing grid isn't just a transmission channel, as the electric grid is. It's also a means of assembling, or integrating, the diverse components of electronic computing into unified and useful services.

Back in the 1990s, Sun Microsystems coined the marketing slogan "The Network Is the Computer." It was catchy but, for most people at the time, meaningless. The network wasn't our computer; the PC on our desk was our computer. Today, Sun's slogan suddenly makes sense. It describes what computing has become, or is becoming, for all of us. The network—the Internet, that is—has become, literally, our computer. The different components that used to be isolated in the closed box of the PC—the hard drive for storing informa-tion, the microchip for processing information, the applications for manipulating information—can now be dispersed throughout the world, integrated through the Internet, and shared by everyone. The World Wide Web has truly turned into the World Wide Computer.

Eric Schmidt, who was still employed by Sun back when it came up with its prophetic slogan, has a different term for the World Wide Computer. He calls it "the computer in the cloud." What he means is that computing, as we experience it today, no longer takes a fixed, concrete form. It occurs in the Internet's ever-shifting "cloud" of data, software, and devices. Our personal computer, not to mention our BlackBerry, our mobile phone, our gaming console, and any other networked gadget we use, is just another molecule of the cloud, another node in the vast computing network. Fulfilling Napster's promise, our PCs have merged with all the other devices on the Internet. That gives each of us using the World Wide Com-puter enormous flexibility in tailoring its workings to our particu-lar needs. We can vary the mix of components—those supplied by utilities and those supplied locally—according to the task we want to accomplish at any given moment.

To put it another way, the World Wide Computer, like any other electronic computer, is programmable. Anyone can write instructions to customize how it works, just as any programmer can write software to govern what a PC does. From the user's perspective, programmability is the most important, the most revolutionary, aspect of utility computing. It's what makes the World Wide Computer a personal computer—even more personal, in fact, than the PC on your desk or in your lap ever was.

To see that programmability in action, you need only look at the controversial online game Second Life. Created by a company named Linden Lab, Second Life is a computer-generated, three-dimensional world. It's populated by players' digitized alter egos, which take the form of computer-generated figures, or "avatars." Although it has some similarities with other massively multiplayer online games like World of Warcraft, Second Life is a very unusual game. There are no rules, no winners or losers. Through their avatars, players simply become citizens, or, as they're usually called, "residents," of a simulated society. Within that society, they can do whatever they want whenever they want. They can stroll down sidewalks, drive down streets, or fly through the air. They can chat with friends or strike up conversations with strangers. They can buy land and build houses. They can shop for clothes and other merchandise. They can take classes or go to business conferences. They can dance in nightclubs and even have sex afterwards.

Second Life is an example of a utility service supplied over the Internet and shared simultaneously by many people. It's very different from traditional computer games, which need to be installed separately on each player's hard drive. But Second Life is also itself a construction of many other utility services. The "computer" that runs Second Life doesn't exist in any one place; it's assembled, on the fly, from various data-storage and data-processing molecules floating around in the global computing cloud. When you join Second Life, you use your Web browser to download a little software program that's installed on your PC's hard drive. Although you initiate the download through Second Life's home page, the program is actually delivered to your computer from storage drives owned and operated by Amazon Web Services. The Second Life home page, the Amazon drives, and the browser running on your PC act in unison to carry out the download. You're not conscious of the intricate exchanges of data going on behind the scenes.

Once installed, the little program stores information about the appearance and the location of your avatar. The program constantly "talks," over the Internet, with the main software Linden Lab uses to generate its online world. That software runs on hundreds of server computers that are housed in two data centers, one in San Francisco and one in Dallas, owned not by Linden Lab but by utility hosting companies. Every server computer contains, in turn, four virtual computers, each of which controls a 16-acre plot of land in Second Life. All the real and virtual computers work in tandem to create the vast world that residents experience as they play the game. In addition, other companies and individuals can write software programs to add new features to Second Life. A resident can, for example, design a virtual necklace that he can then sell to other residents. And he can create a promotional video loop, running on his own computer at home, that can be projected onto the wall of the store he constructs to sell the necklaces. Linden Lab has programmed the World Wide Computer to combine all of these many pieces of software and hardware into the cohesive game that is Second Life.

For a business example, consider how Salesforce.com delivers its account-management service. As with Second Life, the basic software runs on a large number of server computers housed in various data centers. Salesforce's clients use the browser software running on their PCs or mobile phones to tap into the account-management program over the Internet. The browser in effect becomes part of the Salesforce application, serving as its user interface.

But that's just the beginning. Salesforce allows software from many other companies to be incorporated into its service. Google's popular mapping service, Google Maps, can be combined with Salesforce's program to generate road maps showing the locations of customers and prospects. Skype's Internet telephone service can also be brought into the application, allowing sales agents to call customers without leaving their browser. These so-called "mash-ups" between different software services happen automatically and, to the customer, invisibly. All the different pieces of software merge into a single application in the user's browser, even though the programs are owned by different companies and are running on computers in many different places.

CERN, the place where the Web itself got its start, is using the World Wide Computer in a particularly creative way. In 2007, the laboratory completed the construction of the biggest scientific instrument on earth, a particle accelerator called the Large Hadron Collider. As the machine was being built, the lab's computer scientists faced a dilemma. They knew that the collider would produce a huge amount of data for analysis—about 15 petabytes a year. (A petabyte equals one million gigabytes.) Over the course of the collider's expected fifteen-year life, some 225 petabytes would thus need to be stored and processed, a task that would require about 100,000 computers. As a government-funded academic organization, CERN simply couldn't afford to buy and maintain that many machines or even to rent the necessary processing power from a utility supplier.

But the scientists realized that the World Wide Computer gave them another option. They didn't need to pay for new computers at all. Instead, they could ask other research institutions, and even private citizens, to donate the spare computing power and storage capacity from their own PCs and servers. Many thousands of computers, tied together by the Internet, would crunch CERN's numbers as if they were a single supercomputer. What's come to be called the CERN Grid "goes well beyond simple communication between computers," says the institute, "and aims ultimately to turn the global network of computers into one vast computational resource." The CERN Grid represents a very different model from that used by most utility computing pioneers. It doesn't involve the assembly of a lot of gear into a central data center. Rather, it draws on machines scattered across the world. What it has in common with other utilities is the centralization of control. CERN uses sophisticated software to make the thousands of machines perform as if they were one machine. Like Linden Lab and Salesforce.com, it programs the World Wide Computer to do its bidding in the most efficient way possible.

As the capacity of the World Wide Computer expands, it will continue to displace private systems as the preferred platform for computing. Businesses will gain new flexibility in assembling computing services to perform custom information-processing jobs. Able to easily program the World Wide Computer in their own ways, they'll no longer be constrained by the limits of their own data centers or the dictates of a few big IT vendors.

Because of computing's modularity, companies will have a wealth of options as they make the leap to the utility age. They'll be able to continue to fulfill some of their computing requirements through their in-house data centers and IT departments, while relying on outside utilities to satisfy other needs. And they'll be able to continually fine-tune the mix as the capabilities of the utilities advance. In contrast to the switch-over to electric utilities, buyers don't face an all-or-nothing choice when it comes to computing. While smaller companies have strong economic incentives to embrace the full utility model quickly, most larger companies will need to carefully balance their past investments in in-house computing with the benefits provided by utilities. They can be expected to pursue a hybrid approach for many years, supplying some hardware and software requirements themselves and purchasing others over the grid. One of the key challenges for corporate IT departments, in fact, lies in making the right decisions about what to hold on to and what to let go.

In the long run, the IT department is unlikely to survive, at least not in its familiar form. It will have little left to do once the bulk of business computing shifts out of private data centers and into "the cloud." Business units and even individual employees will be able to control the processing of information directly, without the need for legions of technical specialists.

WHERE THINGS GET really interesting, though, is in the home. The ability to program the World Wide Computer isn't limited to corporations and other large organizations. It's available to anyone with a computer hooked up to the Internet. Our houses, like our workplaces, are all becoming part of the computing cloud. Each of us now has a supercomputer, with a virtually unlimited store of data and software, at our beck and call.

Many people are already programming the World Wide Computer, often without knowing it. A simple example will help show how. A man I'll call Geoff is an aficionado of vintage Mustangs— he owns two and has helped restore many others—and a few years back he decided to share his passion with others by putting together a Web site. He registered a domain name, set up a server in a spare room in his home, signed up for an account with a local Internet service provider, and bought an expensive Web-design program called Dreamweaver. His site was a bare-bones production—pages of text interspersed with some photographs, a couple of tables, and a handful of links to other Mustang-related sites—but it took him a long time to get it looking good and operating correctly. Because changing or adding content was a hassle, he found that he rarely updated the pages. Not surprisingly, the site drew few visitors. After a few months, he got bored and abandoned his creation.

Recently, though, Geoff decided to give it another shot. But instead of building a traditional site, he started a blog. Launching it was a snap. Using the browser on his PC, he signed up for an account at the blog-publishing site WordPress. His blog was automatically set up on one of WordPress's servers and assigned its own Web address. Geoff writes his blog entries in a browser window using WordPress's software, which is running on computers owned by Automattic, the company that operates the WordPress service. Every time Geoff completes an entry, he clicks a "publish" button in his browser, and the software saves the entry on WordPress's computers, formats it according to Geoff's instructions, and publishes it on his blog.

Geoff didn't want his blog to be limited to text, though. Using his cameraphone, he had recently made a short video of a big Mustang rally, and he wanted his blog's visitors to be able to watch it. So he transferred the video onto his PC and, again using his browser, uploaded a copy to the YouTube video-hosting service. YouTube translated the file into a format viewable by any computer, and it provided Geoff with a simple cut-and-paste code for adding the video to his blog. Although the video is stored on YouTube's computers, it plays through a window on Geoff's site.

Geoff had also taken some photographs of vintage Mustangs with his digital camera. Wanting to share them as well, he uploaded copies to the Flickr photo-sharing site. He noticed, however, that the colors looked washed-out on-screen. To tweak them, he went to another site, called Phixr, and launched its online photo-editing software. His photos were transferred from Flickr to Phixr automatically, and he used the Phixr tools to boost their color saturation and make a few other adjustments. He saved the changes, sending the enhanced photos back to Flickr. Flickr provided Geoff with another simple code that let him add a photo frame to his blog. The Flickr service feeds a new picture into the frame every few seconds. As with the YouTube video, the photos remain stored on Flickr's computers, though they appear on Geoff's pages. Geoff also noticed that a lot of other people had uploaded photos of old Mustangs to the Flickr site. He instructed Flickr to randomly insert some of those photos into his blog as well.

But he wasn't done yet. A fan of sixties surf rock, Geoff decided he'd like to let visitors see what music he'd been listening to recently. So he signed up for an account with Last.fm, an online service that monitors the songs its members play on their computers and creates a customized online radio station tailored to each member's tastes. Geoff instructed Last.fm to keep a Top 10 list of his most-played songs and to show the list in a box, or "widget," in a column at the side of his blog. Last.fm updates the list every time Geoff plays a new song.

Geoff also wanted to let his readers know who else was reading his blog. He signed up with MyBlogLog for a service that keeps track of his blog's visitors and lists their names—and even their pictures—in another widget. Finally, he wanted to allow his readers to subscribe to his writings. So he set up an account with Feedburner, which provided him with a "subscribe" button to add to his blog. Using the syndication technology knows as RSS, Feedburner alerts subscribers whenever Geoff posts a new article, and it lets Geoff know how many people have signed up.

Geoff's work, which only took him a few hours, gives a sense of how simple it's become to draw data and services from various utility suppliers and combine them onto a single Web page. What's remarkable is that he didn't need to install any software or store any data on his PC—other than, temporarily, the original video and photo files. The various software applications, and all the data, reside on the utility companies' systems. Using simple tools, he programmed all these far-flung machines to create a multimedia experience for his readers. What's even more remarkable is that Geoff didn't pay anything for the software, the storage, the computing power, or the bandwidth through which all the data travels. Everything was free.

In fact, Geoff has even been able to make a little money from his blog by opening up an account with Google's AdSense service. Google automatically places text advertisements on his pages, gearing the ads to the interests of Mustang lovers. Any time a reader clicks on an ad, Google shares the advertising revenue with Geoff. While he was at it, Geoff also signed up for a free account with Google Analytics, which monitors his traffic and provides him with detailed reports on who's visiting his blog, what pages they're looking at, and how long they're spending on each page. Geoff is tapping into Google's vast data centers and enormously complex algorithms—again at no cost to him.

The tools for programming the World Wide Computer are in an early stage of development. While it wasn't difficult for Geoff to construct his blog, he did have to go to many different sites and copy codes manually. In the future, programming tools will get easier to use even as they become more powerful—that's what always happens with software. Yahoo gave us a preview of what's to come when in 2007 it introduced its Pipes programming service. Pipes allows anyone to create a custom Web service by combining and filtering the contents of different Internet databases from a single browser window. As with 3Tera's software for building computer systems, it's all done by dragging and dropping icons. Geoff, for instance, could use Pipes to construct a service that monitors his favorite online information sources—newspapers, magazines, blogs, wikis—and spots any new mention of Mustangs. Every morning, the service would publish a fresh list of headlines on Geoff's blog, with links to the full stories.

IN THE EARLY decades of the twentieth century, as punch-card tabulators and other computing machines gained sophistication, mathematicians and businessmen began to realize that, in the words of one historian, "information is a commodity that can be processed by a machine." Although it now sounds obvious, it was a revolutionary insight, one that fueled the growth and set the course of the entire computer industry, particularly the software end of it, and that is now transforming many other industries and reshaping much of the world's economy. As the price of computing and bandwidth has plunged, it has become economical to transform more and more physical objects into purely digital goods, processing them with computers and transporting and trading them over networks.

We see this phenomenon at work in Geoff's blog, which combines a variety of elements—text, photographs, video, software, music, advertisements—that were traditionally delivered in physical form. But those are just the tip of the iceberg. Many other products are shedding their physical embodiments and turning into pure information, from money to plane tickets to newspapers to X-rays to blueprints to greeting cards to three-dimensional models. What's happening to goods is also happening to places. Many of the everyday interactions that used to have to take place in physical spaces—bank branches, business offices, schools, stores, libraries, theaters, even playgrounds—can now take place more efficiently in virtual spaces.

The melding of the world of real things and places with the world of simulated things and places will only accelerate as the World Wide Computer becomes more powerful and as more devices are hooked up to it. Joint ventures between technology companies and automakers—Google has teamed up with Volkswagen, and Microsoft is working with Ford—promise to turn cars into what one reporter calls "hi-tech computers on wheels." Information and advertisements will flow from the Internet to dashboard displays. Mobile phones, too, are turning into powerful handheld computers—Apple's multipurpose iPhone is a much-discussed example— and Internet service is beginning to be supplied on planes, trains, and ships. Researchers are also making strides in "pervasive computing"—the use of tiny networked sensors to monitor buildings and other physical spaces. The government-funded Center for Embedded Networked Sensing has already covered its own headquarters, at the University of California at Los Angeles, with miniature cameras and microphones connected wirelessly to computers, and it is deploying similar systems at other test sites around the world. The center says its research "has resulted in several new classes of [sensing] systems that can be rapidly distributed in the environment to reveal phenomena with unprecedented detail."

Soon, the World Wide Computer will know where we are and what we're doing at almost every instant of the day. We will exist simultaneously in the real world and in a computer-generated world. In programming the World Wide Computer, we will be programming our lives. Second Life may be only a game, but its central conceit—that we can separate ourselves from our bodies and exist as avatars in a digitized landscape—is more than an amusement. It's a metaphor for our future.

THE DOTCOM CRASH that wiped out so many companies during 2000 and 2001 dampened the optimism surrounding the Internet, but only briefly. Over the last few years, as the computing capacity of the Net exploded and the idea that we'd entered a new Internet era—Web 2.0— took hold, we've seen a resurgence of utopianism. In "We Are the Web," a widely read article that appeared in *Wired* in 2005, Kevin Kelly, a long-time associate of Stewart Brand, described the future in explicitly transcendental terms, claiming to see in the World Wide Computer the seeds of a communal, superhuman intelligence. The Internet, he wrote, is turning into a "megacomputer," a "gargantuan Machine" that "will evolve into an integral extension not only of our senses and bodies but our minds." This machine will become "a collaborative interface for our civilization, a sensing, cognitive device with power that exceed[s] any previous invention." It is providing, he wrote, echoing John Perry Barlow, "a new mind for an old species." In the end, "we will live inside this thing."

Kelly is right about one thing at least. We *are* coming to live inside the World Wide Computer. It's becoming the default forum for many of our commercial and personal relationships, the medium of choice for storing and exchanging information in all its forms, the preferred means of entertaining, informing, and expressing ourselves. The number of hours we spend online every week has been rising steadily for years, and as we've switched from dial-up to broadband connections our reliance on the Web has expanded greatly. For growing numbers of us, in fact, the virtual is becoming as real as the physical. According to a 2007 study by the Annenberg School for Communication's Center for the Digital Future, nearly half of the people who have joined online communities "say they 'feel as strongly' about their virtual community as they do about their real-world communities." The center's director, Jeffrey Cole, described the Internet as "a comprehensive tool that Americans are using to touch the world."

But as we consider what kind of world we'll be touching, we should be wary of the stirring words dispensed by Kelly and the other techno-utopians. Although, as we saw with electrification, optimism is a natural response to the arrival of a powerful and mysterious new technology, it can blind us to more troubling portents. "The simple faith in progress," wrote Norbert Wiener, one of the great theoreticians of information processing, "is not a conviction belonging to strength, but one belonging to acquiescence and hence to weakness." As we'll see, there is reason to believe that our cybernetic meadow may be something less than a new Eden.

CHAPTER 7

From the Many to the Few

I T WAS AN ODD moment in the history of modern business. Usually when one company acquires another, the deal is announced in a carefully staged and scripted event. Reporters are shepherded into a theater or a hotel ballroom, where the chief executives of the two firms stand together at a podium. They speak in general terms of the fabulous prospects for the new company, touting the financial and organizational "synergies" that the combination will provide, and they stress that the deal should be viewed as "a merger of equals" between two businesses with rich and distinctive histories and cultures. There's little spontaneity to the proceedings. The words tend to be as interchangeable as the CEOs' ties.

But when, on October 9, 2006, Google purchased the fledgling Internet video network YouTube, tradition went out the window. Within hours of the deal's completion, a two-minute video appeared on YouTube featuring the company's twentysomething founders Chad Hurley and Steve Chen. Shot on a sidewalk with a handheld camcorder, the video had the cheap, do-it-yourself feel typical of the amateur productions uploaded to the site. For the first few seconds of the clip, the newly wealthy duo, who appeared to have gone without sleep for days and without sun for months, managed to keep their euphoria under control as they searched for the right words to explain the acquisition to what they referred to, repeatedly, as "the YouTube community."

"Hi, YouTube," Hurley began. "This is Chad and Steve. We just want to say thank you. Today, we have some exciting news for you. We've been acquired by Google."

"Yeah, thanks," chimed in Chen, leaning nervously toward the camera. "Thanks to every one of you guys that have been contributing to YouTube, to the community. We wouldn't be anywhere close to where we are without the help of this community."

Struggling to keep a straight face, Hurley continued in a tone of strained seriousness: "We're going to stay committed to developing the best service for you—you know, developing the most innovative service and tools and technologies so you can keep having fun on our site."

But a minute into the video, all pretense of gravity gave way, and the announcement dissolved into a giddy slapstick routine, with the tall, gaunt Hurley playing Stan Laurel to the shorter, round-faced Chen's Oliver Hardy.

"This is great," Hurley said, breaking into a wide grin. "Two kings have gotten together, and we're going to be able to provide you with an even better service."

Chen broke into laughter at the mention of "two kings." He ducked off camera, stumbling down the sidewalk.

"Two kings," repeated Hurley, corralling his partner by grabbing him around the shoulders.

"Get your hand off me, king," said Chen, still laughing.

In a vain attempt to get the founders to straighten up, the cameraman shouted out a question: "What does it mean for the user community?"

"Two kings have gotten together," replied Hurley. "The king of search and the king of video have gotten together. We're going to have it our way. Salt and pepper." Chen doubled over, and Hurley ended the video by drawing a finger across his throat. "We can't do this," he said. "Cut."

The video proved a big hit on YouTube, shooting to the top of the site's most-watched list. Within a month, it had been viewed 2 million times and had inspired a slew of parodies filmed by YouTube members and dutifully uploaded to the site. But though the subversive frivolity of Hurley and Chen's acquisition announcement was remarkable in itself, behind it lay a much deeper break with the past. In the rise of YouTube we see a microcosm of the strange new world of online business. The company's success reveals much about the changing economics of computing and how they're affecting commerce, employment, and even the distribution of wealth.

CHAD HURLEY AND Steve Chen, together with a third friend, Yawad Karem, had come up with the idea of launching an easy-to-use video-sharing service after a dinner party in early 2005. They chose the name YouTube during a brainstorming session on Valentine's Day. Over the next few months, they designed and wrote the code for their site in the garage of the Silicon Valley house that Hurley had purchased with an earlier dotcom windfall. After successfully testing the service in May 2005 by broadcasting a video of Chen's cat playing with a string, they received $3.5 million in funding from a venture capital firm, enough to cover their modest startup costs. In December, the YouTube site officially opened for business, and it rapidly attracted an audience of people looking for a simple—and free—way to store, share, and view short, home-made videos (not to mention thousands of clips illegally copied from copyrighted films, TV shows, and music videos). It was just ten months later that Hurley and Chen sold the site to Google for a staggering $1.65 billion, bringing each of them a windfall worth about a third of a billion dollars.

At the time it was bought, YouTube had just sixty employees. They worked above a pizza joint in San Mateo, California, crammed into a single small office furnished with Ikea desks and chairs, a rubber chicken hanging from the bare metal rafters. (The company's servers were maintained off-site, in data centers run by utility hosting firms.) Despite its minuscule size and lack of formal organization, the staff was able to run one of the most popular and fastest-growing sites on the Internet. Every day, people all over the world watched more than 100 million YouTube video clips and uploaded some 65,000 new videos to the site. And those numbers were growing at an exponential pace, which explained why Google was willing to spend so much to buy the young company. At the $1.65 billion sale price, each YouTube employee represented $27.5 million in market value. Compare that to a traditional, and fabulously profitable, software company like Microsoft, which has 70,000 employees, each representing $4 million in market value. Or compare it to a traditional media and entertainment company like Walt Disney, with 133,000 employees, each representing $500,000 in value.

The abundance of cheap processing power, storage capacity, and communication bandwidth is what made it possible for YouTube to build a very large business very quickly with very few people. And YouTube's experience is far from unique. Many new companies are using the utility computing grid to create burgeoning enterprises with hardly any employees. A year before Google bought YouTube, another Internet giant, eBay, purchased the Internet telephone company Skype for $2.1 billion. Founded just two years earlier by a pair of Scandinavian entrepreneurs, Skype had signed up 53 million customers—more than twice the number of phone customers served by venerable British Telecom—and was attracting 150,000 new subscribers every day. Morgan Stanley said the company's service "may be the fastest-growing product in history." Yet Skype employed just 200 people, about 90,000 fewer than British Telecom employed in the United Kingdom alone. A columnist for a business newspaper in Estonia, where Skype has its largest office, noted that, based on the company's sale price, "one Skype employee is worth more than the Viisnurk wood-processing company and 1.5 employees equal the entire value of the Kalev candy factory."

The classified ad site Craigslist is another example. It was created in 1995 by a software designer named Craig Newmark as an online bulletin board where people could post information about upcoming events in and around San Francisco. After Newmark incorporated his company in 1999, the site expanded rapidly. By the end of 2006, it had bulletin boards for more than 300 cities around the world, each offering a broad mix of for-sale, help-wanted, and personal ads, as well as notices of events and other local activities. More than 10 million visitors were looking at some 5 billion pages on the site every month, making Craigslist one of the most popular destinations on the Web. Yet the entire operation was being run by just twenty-two people.

Perhaps most remarkable of all is PlentyOfFish, an online service that helps people find dates. Launched in Canada in 2003, the site experienced explosive growth. By late 2006, some 300,000 people were logging on to the service every day, and they were looking at about 600 million pages a month. PlentyOfFish had become the largest dating site in Canada and one of the ten biggest in both the United States and the United Kingdom. So how many people does this booming business employ? Precisely one: its founder, Markus Frind. Frind designed and wrote all the code for the site, and he uses Google's automated AdSense service to place advertisements on it, reportedly earning the twenty-eight-year-old entrepreneur as much as $10,000 a day. Entirely computerized, the operation essentially runs itself. In a posting on his blog in June 2006, Frind wrote, "It amazes me to think that my 1 person company is able to compete at a level where all the competitors have . . . 300+ full time staff. I feel like I am ahead of my time, and when I look around at the companies out there today I have a hard time seeing them existing in a few years."

Companies like YouTube, Skype, Craigslist, and PlentyOfFish can grow so quickly with so few workers because their businesses are constructed almost entirely out of software code. Their products are virtual, residing in computer databases or flying across the Internet as strings of digits. The price of creating a fresh copy of their product and distributing it to a new customer anywhere in the world is essentially zero, so the companies can expand without hiring many additional employees. By relying on the public Internet as their distribution channel, moreover, they can avoid many of the capital investments traditional companies have to make. YouTube doesn't have to build production studios or transmission towers. Skype doesn't have to string miles of cable between telephone poles. Craigslist doesn't have to buy printing presses, ink, and paper. PlentyOfFish doesn't have to open offices. Although they have to pay modest fees for the bandwidth they use, they get something of a free ride on the fiber-optic cables that others paid for during the dotcom boom.

All of these businesses demonstrate an unusual sort of economic behavior that economists call "increasing returns to scale." What it means, simply, is that the more products they sell, the more profitable they become. That's a very different dynamic from the one that prevails in the industrial world, where businesses are subject to diminishing returns to scale. As a producer of physical goods increases its output, it sooner or later has to begin paying more for its inputs—for the raw materials, components, supplies, real estate, and workers that it needs to make and sell its products. It can offset these higher input costs by achieving economies of scale—by using fewer inputs to make each additional product—but eventually the higher costs overwhelm the scale economies, and the company's profits, or returns, begin to shrink. The law of diminishing returns in effect sets limits on the size of companies, or at least on the size of their profits.

Until recently, most information goods were also subject to diminishing returns because they had to be distributed in physical form. Words had to be printed on paper, moving pictures had to be captured on film, software code had to be etched onto disks. But because the Internet frees information goods from their physical form, turning them into entirely intangible strings of ones and zeroes, it also frees them from the law of diminishing returns. A digital good can be replicated endlessly for essentially no cost—its producer does not have to increase its purchases of inputs as its business expands. Moreover, through a phenomenon called the network effect, digital goods often become more valuable as more people use them. Every new member that signs up for Skype, puts an ad on Craigslist, or posts a profile on PlentyOfFish increases the value of the service to every other member. Returns keep growing as sales or use expands—without limit.

The unique economics of running a business over the computing grid, combined with the global reach of Web sites, allow Internet-based companies to pursue a business strategy that would have been unthinkable just a few years ago: they can give their core products away for free. YouTube charges its users nothing to store or watch a video; it makes its money through advertising and sponsorships. PlentyOfFish also depends on ads for its income, letting people post and view profiles without charge. Skype allows subscribers to make unlimited phone calls to other subscribers over the Internet—gratis—and charges just a few cents a minute for calls over traditional telephone lines. Craigslist's Newmark doesn't even seem to be interested in having his company make a lot of money. He imposes fees on only a few types of ads—commercial real-estate and job listings, mainly—and gives everything else away as a community service.

THE NEW ECONOMICS of doing business online are a boon to consumers. What used to be expensive—everything from international phone calls to classified ads to video transmissions—can now be had for free. But there's another side to the robotic efficiency, global scope, and increasing returns enjoyed by the new Internet companies. Each of these companies competes, after all, with old-line firms that have long employed and paid decent wages to many people. YouTube fights for viewers with television stations and movie theaters. Skype battles landline and mobile telephone companies for callers. Many of the classified ads that run on Craigslist would have otherwise appeared in local newspapers. Dozens of national and local dating services vie for customers with sites like PlentyOfFish. Given the economic advantages of online firms—advantages that will grow as the maturation of utility computing drives the costs of data processing and communication even lower—traditional firms may have no choice but to refashion their own businesses along similar lines, firing many millions of employees in the process.

We already see signs of the thinning out of the professional workforce in some information industries. As newspapers have lost readers and advertising revenues to Web sites, for instance, they've been forced to lay off reporters and other professionals. A study by the American Society of Newspaper Editors found that between 2001 and 2005 the newsroom staff of US papers declined by 4 percent, with a net loss of 1,000 reporters, 1,000 editors, and 300 photographers and artists. "Web 2.0 and the Net in general have been disasters for my profession," says Philip Dawdy, an award-winning journalist for *Seattle Weekly*. "Newspapers are dying. Talented people are being forced into public relations work."

In early 2007, the US Department of Labor released a revealing set of statistics on the publishing and broadcasting business as a whole. Employment in the industry had fallen by 13 percent in the six years since 2001, with nearly 150,000 jobs lost. These were years when many media companies, as well as their customers and advertisers, had been shifting from physical media to the Internet. Yet the report revealed that there had been no growth in Internet publishing and broadcasting jobs. In fact, online employment had actually dropped 29 percent, from 51,500 to 36,600, during the period. "The Internet is the wave of the future," commented *New York Times* economics writer Floyd Norris. "Just don't try to get a job in it."

Markus Frind's prediction that many traditional businesses will find it impossible to compete against extraordinarily lean Web operations may be coming to pass. As a result, we could well see a hollowing-out of many sectors of the economy, as computers and software displace workers on a large scale. Anyone employed by a business whose product or service can be distributed in digital form may be at risk, and the number of companies that fit that category is growing every year.

Of course, the displacement of workers by computers is nothing new, and it would normally be welcomed as a sign of a healthy economy. Improving labor productivity is what attracted companies to computers in the first place, after all. Computerization, like electrification before it, simply continues the centuries-long trend of substituting machines for workers. As three scholars, David Autor, Frank Levy, and Richard Murnane, explain in an influential article in the *Quarterly Journal of Economics,* "Substitution of machinery for repetitive human labor has been a thrust of technological change throughout the Industrial Revolution. By increasing the feasibility of machine substitution for repetitive human tasks, computerization furthers—and perhaps accelerates—this long-prevailing trend." But computerization, they also point out, brings "a qualitative enlargement in the set of tasks that machines can perform. Because computers can perform symbolic processing—storing, retrieving, and acting upon information—they augment or supplant human cognition in a large set of information-processing tasks that historically were not amenable to mechanization." Computerization extends the replacement of workers by machines from the blue-collar to the white-collar world.

Whereas industrialization in general and electrification in particular created many new office jobs even as they made factories more efficient, computerization is not creating a broad new class of jobs to take the place of those it destroys. As Autor, Levy, and Murnane write, computerization "marks an important reversal. Previous generations of high technology capital sharply increased demand for human input of routine information-processing tasks, as seen in the rapid rise of the clerking occupation in the nineteenth century. Like these technologies, computerization augments demand for clerical and information-processing tasks. But in contrast to [its] predecessors, it permits these tasks to be automated." Computerization creates new work, but it's work that can be done by machines. People aren't necessary.

That doesn't mean that computers can take over all the jobs traditionally done by white-collar workers. As the scholars note, "Tasks demanding flexibility, creativity, generalized problem-solving and complex communications—what we call nonroutine cognitive tasks—do not (yet) lend themselves to computerization." That parenthetical "yet," though, should give us pause. As the power and use-fulness of networked computers have advanced during the few years since they wrote their paper, we've seen not only the expansion of software's capabilities but the flowering of a new phenomenon that is further reducing companies' need for workers. Commonly termed "social production," the phenomenon is reshaping the economics of the media, entertainment, and software industries, among others. In essence, it allows many of those "nonroutine cognitive tasks" that require "flexibility, creativity, generalized problem-solving and complex communications" to be carried out for free—not by computers on the network but by people on the network.

Look more closely at YouTube. It doesn't pay a cent for the hundreds of thousands of videos it broadcasts. All the production costs are shouldered by the users of the service. They're the directors, producers, writers, and actors, and by uploading their work to the YouTube site they're in effect donating their labor to the company. Such contributions of "user-generated content," as it's called, have become commonplace on the Internet, and they're providing the raw material for many Web businesses. Millions of people freely share their words and ideas through blogs and blog comments, which are often collected and syndicated by corporations. The contributors to open-source software projects, too, donate their labor, even though the products of their work are often commercialized by for-profit companies like IBM, Red Hat, and Oracle. The popular online encyclopedia Wikipedia is written and edited by volunteers. Yelp, a group of city sites, relies on reviews of restaurants, shops, and other local attractions contributed by members. The news agency Reuters syndicates photos and videos submitted by amateurs, some of whom are paid a small fee but most of whom get nothing. Social networking sites like MySpace and Facebook, and dating sites like PlentyOfFish, are essentially agglomerations of the creative, unpaid contributions of their members. In a twist on the old agricultural practice of sharecropping, the site owners provide the digital real estate and tools, let the members do all the work, and then harvest the economic rewards.

The free labor is not limited to the production of creative works. Popular news-filtering sites like Digg and Reddit rank stories based on the votes of millions of members, obviating the need for editors. The photographs uploaded to Flickr are all sorted on the basis of labels, or "tags," placed on them by the site's users. Del.icio.us offers a similar service for tagging Web pages, and Freebase offers one for tagging information of all sorts. Much of the processing power required to run the Skype network is "borrowed" from the microprocessors inside users' PCs, dramatically reducing the number of computers Skype has to buy. Even the search services provided by companies like Google and Yahoo are essentially constructed from the contributions of the makers and viewers of Web sites. The search firms don't need to hire the analysts and librarians who have traditionally categorized and filtered information for the benefit of others. More and more companies are figuring out ways to harness the power of free labor. Even police forces are getting into the act. In 2006, Texas marshals set up webcams along the border with Mexico and began streaming the video feeds over the Internet. People all around the world can now watch for illegal immigrants, clicking a button to alert the police to any suspicious activity. It's law enforcement on the cheap.

Why do people contribute their labor in this way? There are several reasons, none of them particularly surprising. In some cases, such as the building of search engines, they contribute without even knowing it. Companies like Google simply track people's everyday behavior online and distill valuable intelligence from the patterns the behavior reveals. No one minds because the resulting products, like search results, are useful. In other cases, people contribute out of their own self-interest. Creating a MySpace or a Facebook page provides a social benefit to many young people, helping them stay in touch with old friends and meet new ones. Tagging photos at Flickr or Web pages at Del.icio.us helps people keep track of words and images that interest them—it serves as a kind of personal filing system for online content. Some sites share a portion of their advertising revenues with contributors (though the sums are usually trivial). In still other cases, there's a competitive or status-seeking element to the donations. Sites like Digg, Yelp, and even Wikipedia have hierarchies of contributors, and the more you contribute, the higher you rise in the hierarchy.

But the biggest reason people contribute to such sites is no different from the reason they pursue hobbies or donate their time to charitable causes or community groups: because they enjoy it. It gives them satisfaction. People naturally like to create things, to show off their creations to others, to talk about themselves and their families, and to be part of communal projects. It's no different on the Internet. Even very early online communities and Web sites made extensive use of free labor. In the 1990s, so many thousands of America Online members were performing unpaid jobs for the company, such as moderating chat rooms, that *Wired* magazine called AOL a "cyber-sweatshop." Much of Amazon.com's early appeal came from the book reviews donated by customers—and the ratings of those reviews submitted by other customers. The uploading of videos, the writing of blogs, the debugging of open-source code, the editing of Wikipedia entries—all are simply new forms of the pastimes or charitable work that people have always engaged in outside their paid jobs.

What has changed, though, is the scope, scale, and sophistication of the contributions—and, equally important, the ability of companies to harness the free labor and turn it into valuable products and services. Ubiquitous, inexpensive computing and data communication, together with ever more advanced software programs, allow individuals to make and share creative works and other information goods in ways that were never possible before, and they also enable thousands or even millions of discrete contributions to be assembled into commercial goods with extraordinary efficiency. In his book *The Wealth of Networks,* Yale law professor Yochai Benkler traces the recent explosion in social production to three technological advances. "First, the physical machinery necessary to participate in information and cultural production is almost universally distributed in the population of the advanced economies," he writes. "Second, the primary raw materials in the information economy, unlike the physical economy, are [freely available] public goods—existing information, knowledge, and culture." Finally, the Internet provides a platform for distributed, modular production that "allows many diversely motivated people to act for a wide range of reasons that, in combination, cohere into new useful information, knowledge, and cultural goods."

All three factors will become even more salient in the years ahead. The World Wide Computer will continue to give individuals new production capabilities, to expand their access to information, and to make it easier to meld their contributions into useful and attractive products and services. Benkler sees a kind of popular revolution in the making, where the means of producing and distributing information goods, formerly controlled by big companies, are put into the hands of the masses. He believes the "networked information economy" marks "a significant inflection point for modern societies," which promises to bring "a quite basic transformation in how we perceive the world around us." By changing "the way we create and exchange information, knowledge, and culture," he writes, "we can make the twenty-first century one that offers individuals greater autonomy, political communities greater democracy, and societies greater opportunities for cultural self-reflection and human connection."

Other writers and scholars have made arguments similar to Benkler's. They see a new and liberating economy emerging—a "gift economy" that, based on sharing rather than selling, exists outside of and even in opposition to the market economy. Although the term "gift economy" dates back at least a quarter century to Lewis Hyde's 1983 book *The Gift: Imagination and the Erotic Life of Property*, its new vogue highlights how extensive free labor and its products have become on the Internet. The gift economy, it's often said, is engendering a richer, more egalitarian culture, while drawing wealth and power away from the corporations and governments that have, allegedly, monopolized the distribution of creative works and other information goods. Richard Barbrook, of the University of Westminster in London, expressed this view well in his 1998 essay "The Hi-Tech Gift Economy." He wrote of Internet users: "Unrestricted by physical distance, they collaborate with each other without the direct mediation of money or politics. Unconcerned about copyright, they give and receive information without thought of payment. In the absence of states or markets to mediate social bonds, network communities are instead formed through the mutual obligations created by gifts of time and ideas."

There's truth in such claims, as anyone looking at the Web today can see. Computers and the Internet have given people powerful new tools for expressing themselves, for distributing their work to broad audiences, and for collaborating to produce various goods. But there's a naïveté, or at least a short-sightedness, to these arguments as well. The utopian rhetoric ignores the fact that the market economy is rapidly subsuming the gift economy. The "gifts of time and ideas" are becoming inputs to the creation of commodities. Whether it's a big company like Rupert Murdoch's News Corporation, which owns MySpace, or a one-man operation like Markus Frind's Plenty OfFish, businesses are using the masses of Internet gift-givers as a global pool of cut-rate labor.

When, in 2005, the Internet giant Yahoo acquired Flickr for a reported $35 million, the larger company freely admitted that it was motivated by the prospect of harvesting all the free labor supplied by Flickr's users. As Yahoo executive Bradley Horowitz told *Newsweek*, "With less than 10 people on the payroll, [Flickr] had millions of users generating content, millions of users organizing that content for them, tens of thousands of users distributing that across the Internet, and thousands of people not on the payroll actually building the thing. That's a neat trick. If we could do that same thing with Yahoo, and take our half-billion user base and achieve the same kind of effect, we knew we were on to something."

As user-generated content continues to be commercialized, it seems likely that the largest threat posed by social production won't be to big corporations but to individual professionals—to the journalists, editors, photographers, researchers, analysts, librarians, and other information workers who can be replaced by, as Horowitz put it, "people not on the payroll." Sion Touhig, a distinguished British photojournalist, points to the "glut of images freely or cheaply available on the Web" in arguing that "the Internet 'economy' has devastated my sector." Why pay a professional to do something that an amateur is happy to do for free?

There have always been volunteers, of course, but unpaid workers are now able to replace paid workers on a scale far beyond what's been possible before. Businesses have even come up with a buzzword for the phenomenon: "crowdsourcing." By putting the means of production into the hands of the masses but withholding from those masses any ownership over the products of their communal work, the World Wide Computer provides an incredibly efficient mechanism for harvesting the economic value of the labor provided by the very many and concentrating it in the hands of the very few. Chad Hurley and Steve Chen had good reason to thank the "YouTube community" so profusely when announcing the Google buyout. It was the members of that community who had, by donating their time and creativity to the site, made the two founders extremely rich young men.

THE CREATION OF the electric grid accelerated the concentration of wealth in large businesses, a trend that had been progressing, if at a slower pace, since the start of the Industrial Revolution. But as the big companies expanded and introduced new categories of consumer goods, they had to hire huge numbers of both skilled and unskilled workers and, following Henry Ford's precedent, pay them good wages. In this way, electrification forced the companies to spread their increasing wealth widely among their employees. As we've seen, that played a decisive role in creating a prosperous middle class—and a more egalitarian American society.

The arrival of the universal computing grid portends a very different kind of economic realignment. Rather than concentrating wealth in the hands of a small number of companies, it may concentrate wealth in the hands of a small number of individuals, eroding the middle class and widening the divide between haves and have-nots. Once again, this would mark not the beginning of a new trend but rather the acceleration of an existing one.

Since the early 1980s, when businesses' investments in computers began to skyrocket, American incomes have become increasingly skewed. As the incomes of the rich have leapt upward, most people's wages have stagnated. In an extensive study of Internal Revenue Service data, the economists Thomas Piketty and Emmanuel Saez found that the share of overall income that went to the richest 10 percent of households remained steady at around 32 percent between the end of the Second World War and 1980. At that point, it began to creep upward, reaching 34 percent in 1985, 39 percent in 1990, 41 percent in 1995, and 43 percent in 2000. After a brief drop following the stock market's retreat in the wake of the dotcom bust, it returned to 43 percent in 2004.

Even more dramatic, though, has been the increasing concentration of income among the wealthiest of the wealthy. The share of total income held by the richest 1 percent of Americans declined from about 11 percent in the late 1940s to approximately 8 percent in the late 1970s. But the trend reversed itself in the 1980s. By 2004, the top 1 percent were making 16 percent of the money. At the very top of the ladder, the concentration of income has been yet more striking. The share of income held by the top 0.1 percent of American families more than tripled between 1980 and 2004, jumping from 2 percent to 7 percent. When the researchers updated their statistics in late 2006, they saw a continuation of the trend, writing that preliminary tax data "suggests that top incomes have continued to surge in 2005."

In another recent study, researchers from Harvard University and the Federal Reserve examined the pay rates of the top three executives at the largest American corporations and compared them to the average compensation earned by workers in general. They found that executive pay remained fairly stable from the end of the Second World War through the 1970s but has shot up in more recent years. In 1970, according to the study, the median senior business executive earned 25 times more than the average US worker. By 2004, the median executive was receiving 104 times the average worker's pay—and the top 10 percent of executives were earning at least 350 times the average. Ajay Kapur, an economist who serves as Citigroup's chief global equity strategist, argues that the United States, along with Canada and the United Kingdom, has become a "plutonomy" where "economic growth is powered by, and largely consumed by, the wealthy few."

Economists continue to debate the causes of the growing inequality in American incomes. Many factors are involved, including the strengthening of international trade, rising corporate profits, tax-rate changes, the declining power of trade unions, and changing social norms. But it seems increasingly clear that computerization has played a central role in the shift, particularly in holding down the incomes of the bulk of Americans. The distinguished Columbia University economist Jagdish Bhagwati argues that computerization is the main cause behind the two-decades-long stagnation of middle-class wages. "There are assembly lines today, but they are without workers," he writes; "they are managed by computers in a glass cage above, with highly skilled engineers in charge." Normally, the introduction of a new labor-saving technology would erode wages only briefly before the resulting boost in productivity pushed them up again. But information technology is different, Bhagwati contends. Unlike earlier technologies that caused "discrete changes," such as the steam engine, the ongoing advances in computer technology offer workers no respite. The displacement of workers "is continuous now," he says. "The pressure on wages becomes relentless."

In a February 2007 speech, Federal Reserve chairman Ben Bernanke backed up Bhagwati's conclusion that information technology is the primary force behind the skewing of incomes. He argued that "the influence of globalization on inequality has been moderate and almost surely less important than the effects of skill-biased technical change." With software automating ever more highly skilled tasks, the number of people who find their jobs at risk is expanding—as newspaper reporters and editors will testify. The effect is amplified by companies' ability to shift knowledge work, the realm of well-paid white-collar workers, across national borders. Since everyone on the Internet has equal access to the World Wide Computer, a worker's location matters much less than it used to. Any job that can be done on a computer, no matter how specialized, has become portable. Even the reading of the X-rays of US patients is today being "offshored" to doctors in India and other countries. The international competition for jobs pushes salaries down in higher-wage countries like the United States, as the global labor market seeks equilibrium. Computerization hence puts many American wage-earners in a double bind: it reduces the demand for their jobs even as it expands the supply of workers ready and able to perform them.

There's a natural tendency, and a natural desire, to see the Internet as a leveling force, one that creates a fairer, more democratic society, where economic opportunities and rewards are spread widely among the many rather than held narrowly by the few. The fact that the World Wide Computer gives people new freedom to distribute their works and ideas to others, with none of the constraints of the physical world, appears to buttress that assumption. But the reality may be very different. In his 2006 book *The Long Tail*, Chris Anderson writes that "millions of ordinary people [now] have the tools and the role models to become amateur producers. Some of them will also have talent and vision. Because the means of production have spread so widely and to so many people, the talented and visionary ones, even if they're a small fraction of the total, are becoming a force to be reckoned with." This is not, as it might first appear, a vision of a world of economic egalitarianism. It's a vision of a world in which more and more of the wealth produced by markets is likely to be funneled to "a small fraction" of particularly talented individuals.

As we saw with electrification, the interplay of technological and economic forces rarely produces the results we at first expect. There are some who remain convinced that computerization, as it continues to gain momentum, will begin to close the wealth gap that up to now it has helped widen. That's the pattern that has occurred in past technological revolutions. But when we take into account the economic forces that the World Wide Computer is unleashing— the spread of the increasing-returns dynamic to more sectors of the economy, the replacement of skilled as well as unskilled workers with software, the global trade in knowledge work, and the ability of companies to aggregate volunteer labor and harvest its economic value—we're left with a prospect that is far from utopian. The erosion of the middle class may well accelerate, as the divide widens between a relatively small group of extraordinarily wealthy people—the digital elite—and a very large set of people who face eroding fortunes. In the YouTube economy, everyone is free to play, but only a few reap the rewards.

| CHAPTER 8 |

The Great Unbundling

ELECTRIFICATION HASTENED the expansion of America's mass culture, giving people a shared set of experiences through popular television shows, radio programs, songs, movies, books and magazines, newspaper stories, and even advertisements. It opened up new nationwide channels for broadcast media and gave rise to large news and entertainment conglomerates that could afford the investments needed to make and distribute creative works. The advantages of physical scale and geographic reach enjoyed by manufacturers like Ford and General Electric also accrued to media companies like CBS, RCA, Metro–Goldwyn–Mayer, and Time Inc. And because the costs of producing mass media were so onerous, the companies had a strong incentive to pitch a small set of products to as large an audience as possible. In many cases, they had little choice but to restrict production. The airwaves could handle only so many TV and radio programs, shops could stock only so many books and records, and movie theaters could screen only so many films.

The nation's mass culture, and the sense of unity that it instilled in a motley population scattered across a vast land, was not, in other words, the expression of an essential quality of the American character. It was a by-product of the economic and technological forces that swept the country at the start of the twentieth century. The Internet, which is becoming not just a universal computer but also a universal medium, unleashes a very different set of forces, and they promise to reshape America's culture once again.

The major constraints on the supply of creative works—high costs and narrow distribution channels—are disappearing. Because most common cultural goods consist of words, images, or sounds, which all can be expressed in digital form, they are becoming as cheap to reproduce and distribute as any other information product. Many of them are also becoming easier to create, thanks to the software and storage services of the World Wide Computer and inexpensive production tools like camcorders, microphones, digital cameras, and scanners. Tasks that once required a lot of money and training, from film developing to video editing to graphic design to sound mixing, can now be done by amateurs in their dens, offices, and schoolrooms. The proliferation of blogs, podcasts, video clips, and MP3s testifies to the new economics of culture creation. And all the new digital products, whether fashioned by professionals or amateurs, can find a place in the online marketplace. The virtual shelves of the Internet can expand to accommodate everything.

The shift from scarcity to abundance in media means that, when it comes to deciding what to read, watch, and listen to, we have far more choices than our parents or grandparents did. We're able to indulge our personal tastes as never before, to design and wrap ourselves in our own private cultures. "Once the most popular fare defined our culture," explains Chris Anderson. "Now a million niches define our culture." The vast array of choices is exciting, and by providing an alternative to the often bland products of the mass media it seems liberating as well. It promises, as Anderson says, to free us from "the tyranny of lowest-common-denominator fare" and establish in its place "a world of infinite variety."

But while it's true that the reduction in production and distribution costs is bringing us many more options, it would be a mistake to leap to the conclusion that nothing will be sacrificed in the process. More choices don't necessarily mean better choices. Many cultural goods remain expensive to create or require the painstaking work of talented professionals, and it's worth considering how the changing economics of media will affect them. Will these goods be able to find a large enough paying audience to underwrite their existence, or will they end up being crowded out of the marketplace by the proliferation of free, easily accessible products? Even though the Internet can in theory accommodate a nearly infinite variety of information goods, that doesn't mean that the market will be able to support all of them. Some of the most cherished creative works may not survive the transition to the Web's teeming bazaar.

THE TENSIONS CREATED by the new economics of production and consumption are visible today in many media, from music to movies. Nowhere, though, have they been so clearly on display, and so unsettling, as in the newspaper business. Long a mainstay of culture, print journalism is going through a wrenching transformation, and its future is in doubt. Over the past two decades, newspaper readership in the United States has plummeted. After peaking in 1984, at 63 million copies, the daily circulation of American papers fell steadily at a rate of about 1 percent a year until 2004, when it hit 55 million. Since then, the pace of the decline has accelerated. Circulation fell by more than 2 percent in 2005 and by about 3 percent in 2006. Many of the country's largest papers have been particularly hard hit. In just the six months between April and September of 2006, the daily circulation of the *Miami Herald* fell 8.8 percent; the

Los Angeles Times, 8.0 percent; the *Boston Globe,* 6.7 percent; the *New York Times,* 3.5 percent; and the *Washington Post,* 3.3 percent. In 1964, 81 percent of American adults read a daily newspaper. In 2006, only 50 percent did. The decline has been sharpest among young adults. Just 36 percent of 18- to 24-year-olds reported reading a daily newspaper in 2006, down from 73 percent in 1970.

There are many reasons for the long-term decline in newspaper readership. But one of the most important factors behind the recent acceleration of the trend is the easy availability of news reports and headlines on the Internet. As broadband connections have become more common, the number of American adults who get news online every day has jumped, from 19 million in March 2000 to 44 million in December 2005, according to the Pew Internet & American Life Project. The shift to online news sources is particularly strong among younger Americans. At the end of 2005, the Web had become a daily source of news for 46 percent of adults under thirty-six years of age who had broadband connections, while only 28 percent of that group reported reading a local newspaper.

The loss of readers means a loss of advertising revenue. As people continue to spend more time online, advertisers have been moving more of their spending to the Web, a trend expected to accelerate in coming years. From 2004 through 2007, newspapers lost an estimated $890 million in ad revenues to the Internet, according to Citibank research. Classified advertising, long a lucrative niche for newspapers, has been particularly hard hit, as companies and home-owners shift to using sites like Craigslist, eBay, and Autotrader to sell cars and other used goods and to list their apartments and houses. In 2006, sales of classified ads by Web sites surpassed those of newspapers for the first time.

Newspaper companies are, naturally, following their readers and advertisers online. They're expanding their Web sites and shifting ever more of their content onto them. After having kept their print and Web units separate for many years, dedicating most of their money and talent to print editions, papers have begun merging the operations, assigning more of their top editors' time to online editions. During 2006 and 2007, the *New York Times, Washington Post,* and *Wall Street Journal* all announced plans to give more emphasis to their Web sites. "For virtually every newspaper," says one industry analyst, "their only growth area is online." Statistics underscore the point. Visits to newspaper Web sites shot up 22 percent in 2006 alone.

But the nature of a newspaper, both as a medium for information and as a business, changes when it loses its physical form and shifts to the Internet. It gets read in a different way, and it makes money in a different way. A print newspaper provides an array of content—local stories, national and international reports, news analyses, editorials and opinion columns, photographs, sports scores, stock tables, TV listings, cartoons, and a variety of classified and display advertising— all bundled together into a single product. People subscribe to the bundle, or buy it at a newsstand, and advertisers pay to catch readers' eyes as they thumb through the pages. The publisher's goal is to make the entire package as attractive as possible to a broad set of readers and advertisers. The newspaper as a whole is what matters, and as a product it's worth more than the sum of its parts.

When a newspaper moves online, the bundle falls apart. Readers don't flip through a mix of stories, advertisements, and other bits of content. They go directly to a particular story that interests them, often ignoring everything else. In many cases, they bypass the newspaper's "front page" altogether, using search engines, feed readers, or headline aggregators like Google News, Digg, and Daylife to leap directly to an individual story. They may not even be aware of which newspaper's site they've arrived at. For the publisher, the newspaper as a whole becomes far less important. What matters are the parts. Each story becomes a separate product standing naked in the marketplace. It lives or dies on its own economic merits.

Because few newspapers, other than specialized ones like the *Wall Street Journal,* are able to charge anything for their online editions, the success of a story as a product is judged by the advertising revenue it generates. Advertisers no longer have to pay to appear in a bundle. Using sophisticated ad placement services like Google AdWords or Yahoo Search Marketing, they can target their ads to the subject matter of an individual story or even to the particular readers it attracts, and they only pay the publisher a fee when a reader views an ad or, as is increasingly the case, clicks on it. Each ad, moreover, carries a different price, depending on how valuable a viewing or a clickthrough is to the advertiser. A pharmaceutical company will pay a lot for every click on an ad for a new drug, for instance, because every new customer it attracts will generate a lot of sales. Since all page views and ad clickthroughs are meticulously tracked, the publisher knows precisely how many times each ad is seen, how many times it is clicked, and the revenue that each view or clickthrough produces.

The most successful articles, in economic terms, are the ones that not only draw a lot of readers but deal with subjects that attract high-priced ads. And the most successful of all are those that attract a lot of readers who are inclined to click on the high-priced ads. An article about new treatments for depression would, for instance, tend to be especially lucrative, since it would attract expensive drug ads and draw a large number of readers who are interested in new depression treatments and hence likely to click on ads for psychiatric drugs. Articles about saving for retirement or buying a new car or putting an addition onto a home would also tend to throw off a large profit, for similar reasons. On the other hand, a long investigative article on government corruption or the resurgence of malaria in Africa would be much less likely to produce substantial ad revenues. Even if it attracts a lot of readers, a long shot in itself, it doesn't cover a subject that advertisers want to be associated with or that would produce a lot of valuable clickthroughs. In general, articles on serious and complex subjects, from politics to wars to international affairs, will fail to generate attractive ad revenues.

Such hard journalism also tends to be expensive to produce. A publisher has to assign talented journalists to a long-term reporting effort, which may or may not end in a story, and has to pay their salaries and benefits during that time. The publisher may also have to shell out for a lot of expensive flights and hotel stays, or even set up an overseas bureau. When bundled into a print edition, hard journalism can add considerably to the overall value of a newspaper. Not least, it can raise the prestige of the paper, making it more attractive to subscribers and advertisers. Online, however, most hard journalism becomes difficult to justify economically. Getting a freelance writer to dash off a review of high-definition television sets—or, better yet, getting readers to contribute their own reviews for free—would produce much more attractive returns.

In a 2005 interview, a reporter for the *Rocky Mountain News* asked Craig Newmark what he'd do if he ran a newspaper that was losing its classified ads to sites like Craigslist. "I'd be moving to the Web faster," he replied, and "hiring more investigative journalists." It's a happy thought, but it ignores the economics of online publishing. As soon as a newspaper is unbundled, an intricate and, until now, largely invisible system of subsidization quickly unravels. Classified ads, for instance, can no longer help to underwrite the salaries of investigative journalists or overseas correspondents. Each piece of content has to compete separately, consuming costs and generating revenues in isolation. So if you're a beleaguered publisher, losing readers and money and facing Wall Street's wrath, what are you going do as you shift your content online? Hire more investigative journalists? Or publish more articles about consumer electronics? It seems clear that as newspapers adapt to the economics of the Web, they are far more likely to continue to fire reporters than hire new ones.

Speaking before the Online Publishing Association in 2006, the head of the *New York Times*'s Web operation, Martin Nisenholtz, summed up the dilemma facing newspapers today. He asked the audience a simple question: "How do we create high-quality content in a world where advertisers want to pay by the click, and consumers don't want to pay at all?" The answer may turn out to be equally simple: we don't. At least one major newspaper, *The Times* of London, admits that it has already begun training its reporters to craft their stories in ways that lead to higher placements in search engines. Jim Warren, the *Chicago Tribune*'s managing editor, says that "you can't really avoid the fact that page views are increasingly the coin of the realm." As long as algorithms determine the distribution of profits, they will also determine what gets published.

The unbundling of content is not unique to newspapers or other print publications. It's a common feature of most online media. Apple's iTunes store has unbundled music, making it easy to buy by the song rather than the album. Digital video recorders like TiVo and pay-per-view cable services are unbundling television, separating the program from the network and its schedule. Video sites like YouTube go even further, letting viewers watch brief clips rather than sitting through entire shows. Amazon.com has announced plans to unbundle books, selling them by the page. Google provides "snippets" of text from published works through its controversial

Book Search service. Podcasting is unbundling radio programs. Wikipedia is unbundling the encyclopedia. The "bundling of the world's computers into a single network," writes Daniel Akst, "is ushering in what may be called the unbundled age."

Economists are quick to applaud the breaking up of media products into their component pieces. In their view, it's how markets should work. Consumers should be able to buy precisely what they want without having to "waste" money on what they don't. The *Wall Street Journal* celebrates the development, saying it heralds a new era in which we'll no longer have "to pay for detritus to get the good stuff." That's true in many cases, but not in all. Creative works are not like other consumer goods, and the economic efficiency that would be welcomed in most markets may have less salutary effects when applied to the building blocks of culture. It's worth remembering, as well, that the Internet is a very unusual marketplace, where information of all sorts tends to be given away and money is made through indirect means like advertising. Once you fragment both the audience and the advertising in such a market, large investments in the production of certain creative works become much harder for businesses to justify.

If the news business is any indication, the "detritus" that ends up being culled from our culture may include products that many of us would define as "the good stuff." What's sacrificed may not be blandness but quality. We may find that the culture of abundance being produced by the World Wide Computer is really just a culture of mediocrity—many miles wide but only a fraction of an inch deep.

IN 1971, THE economist Thomas Schelling performed a simple experiment that had a very surprising result. He was curious about the persistence of extreme racial segregation in the country. He knew that most Americans are not racists or bigots, that we're generally happy to be around people who don't look or think the same way we do. At the same time, he knew that we're not entirely unbiased in the choices we make about where we live and whom we associate with. Most of us have a preference, if only a slight one, to be near at least some people who are similar to ourselves. We don't want to be the only black person or white person, or the only liberal or conservative, on the block. Schelling wondered whether such small biases might, over the long run, influence the makeup of neighborhoods.

He began his experiment by drawing a grid of squares on a piece of paper, creating a pattern resembling an oversized checkerboard. Each square represented a house lot. He then randomly placed a black or a white marker in some of the squares. Each marker represented either a black or a white family. Schelling assumed that each family desired to live in a racially mixed neighborhood, and that's exactly what his grid showed at the start: the white families and the black families were spread across the grid in an entirely arbitrary fashion. It was a fully integrated community. He then made a further assumption: that each family would prefer to have some nearby neighbors of the same color as themselves. If the percentage of neighbors of the same color fell beneath 50 percent, a family would have a tendency to move to a new house.

On the basis of that one simple rule, Schelling began shifting the markers around the grid. If a black marker's neighbors were more than 50 percent white or if a white marker's neighbors were more than 50 percent black, he'd move the marker to the closest unoccupied square. He continued moving the pieces until no marker had neighbors that were more than 50 percent of the other color. At that point, to Schelling's astonishment, the grid had become completely segregated. All the white markers had congregated in one area, and all the black markers had congregated in another. A modest, natural preference to live near at least a few people sharing a similar characteristic had the effect, as it influenced many individual decisions, of producing a sharp divide in the population. "In some cases," Schelling explained, "small incentives, almost imperceptible differentials, can lead to strikingly polarized results."

It was a profound insight, one that, years later, would be cited by the Royal Swedish Society of Sciences when it presented Schelling with the 2005 Nobel Prize in Economics. Mark Buchanan, in his book *Nexus*, summarized the broader lesson of the experiment well: "Social realities are fashioned not only by the desires of people but also by the action of blind and more or less mechanical forces—in this case forces that can amplify slight and seemingly harmless personal preferences into dramatic and troubling consequences."

Just as it's assumed that the Internet will promote a rich and diverse culture, it's also assumed that it will bring people into greater harmony, that it will breed greater understanding and help ameliorate political and social tensions. On the face of it, that expectation seems entirely reasonable. After all, the Internet erases the physical boundaries that separate us, allows the free exchange of information about the thoughts and lives of others, and provides an egalitarian forum in which all views can get an airing. The optimistic view was perhaps best expressed by Nicholas Negroponte, the head of MIT's Media Lab, in his 1995 bestseller *Being Digital*. "While the politicians struggle with the baggage of history, a new generation is emerging from the digital landscape free of many of the old prejudices," he wrote. "Digital technology can be a natural force drawing people into greater world harmony."

But Schelling's simple experiment calls this view into question. Not only will the process of polarization tend to play out in virtual communities in the same way it does in neighborhoods, but it seems likely to proceed much more quickly online. In the real world, with its mortgages and schools and jobs, the mechanical forces of segregation move slowly. There are brakes on the speed with which we pull up stakes and move to a new house. Internet communities have no such constraints. Making a community-defining decision is as simple as clicking a link. Every time we subscribe to a blog, add a friend to our social network, categorize an email message as spam, or even choose a site from a list of search results, we are making a decision that defines, in some small way, whom we associate with and what information we pay attention to. Given the presence of even a slight bias to be connected with people similar to ourselves—ones who share, say, our political views or our cultural preferences—we would, like Schelling's hypothetical homeowners, end up in ever more polarized and homogeneous communities. We would click our way to a fractured society.

Greatly amplifying the polarization effect are the personalization algorithms and filters that are so common on the Internet and that often work without our permission or even our knowledge. Every time we buy a book at Amazon or rent a movie from Netflix or view a news story at Reddit, the site stores information about our choice in a personal profile and uses it to recommend similar products or stories in the future. The effect, in the short run, can be to expose us to items we wouldn't otherwise have come across. But over the long run, the more we click, the more we tend to narrow the information we see.

As the dominant search engine, Google wields enormous influence over the information people find on the Web, and it has been particularly aggressive in engineering the personalization of content. In 2005, it began testing a personalized search service that "orders your search results based on your past searches, as well as the search results and news headlines you've clicked on." In 2007, it quietly made personalized search the default setting for anyone with a Gmail address or other Google account. (The company's top three competitors, Yahoo, Microsoft, and Ask, also have personalized search tools in the works.) Google scientists have even developed an "audio-fingerprinting" system that can use the microphone in your computer to monitor the "ambient audio" in your room and use it for personalization purposes. If you have your television on, the system can identify the program you're watching by recording a sample of its audio signal and comparing it to an "audio database" stored in a Google data center. The company could then feed you stories or ads keyed to your favorite shows.

Google has said that its goal is to store "100% of a user's data" inside its utility, allowing it to achieve what it calls "transparent personalization." At that point, it would be able to automatically choose which information to show you, and which to withhold, without having to wait for you to ask. It says, for example, that people "should not have to tell us which [news] feeds they want to subscribe to. We should be able to determine this implicitly."

A company run by mathematicians and engineers, Google seems oblivious to the possible social costs of transparent personalization. Its interest, as its CEO has said, lies in "using technology to solve problems that have never been solved before"—and personalization is just one of those problems. But, of course, Google and its competitors are not imposing personalization on us against our will. They're just responding to our desires. We welcome personalization tools and algorithms because they let us get precisely what we want when we want it, with a minimum of fuss. By filtering out "the detritus" and delivering only "the good stuff," they allow us to combine fragments of unbundled information into new bundles, tailor-made for audiences of one. They impose homogeneity on the Internet's wild heterogeneity. As the tools and algorithms become more sophisticated and our online profiles more refined, the Internet will act increasingly as an incredibly sensitive feedback loop, constantly playing back to us, in amplified form, our existing preferences.

In "Global Village or Cyber-Balkans?," an article that appeared in the journal *Management Science* in 2005, Eric Brynjolfsson, of MIT, and Marshall Van Alstyne, of Boston University, describe the results of a mathematical model they constructed to measure how individuals' choices influence the makeup of online communities. "Although the conventional wisdom has stressed the integrating effects of [Internet] technology," they write, in introducing their research, "we examine critically the claim that a global village is the inexorable result of increased connectivity."

They note that, because there are limits to how much information we can process and how many people we can communicate with (we have "bounded rationality," to use the academic jargon), we naturally have a strong desire to use filters to screen the ideas we're exposed to and the people we associate with. As the filters become more finely tuned, we can focus our attention—and structure our communities—with ever greater precision. Schelling's work shows that this process naturally breeds homogeneity in the real world, and Brynjolfsson and Van Alstyne's model confirms that the effects could be even more extreme in the virtual world. "Our analysis," they write, "suggests that automatic search tools and filters that route communications among people based on their views, reputations, past statements or personal characteristics are not necessarily benign in their effects." Shaped by such tools, online communities could actually end up being less diverse than communities defined by physical proximity. Diversity in the physical world "can give way to virtual homogeneity as specialized communities coalesce across geographic boundaries."

They stress that such "balkanization" is not the only possible result of filtering. In theory, "preferences for broader knowledge, or even randomized information, can also be indulged." In reality, though, our slight bias in favor of similarity over dissimilarity is difficult, if not impossible, to eradicate. It's part of human nature. Not surprisingly, then, Brynjolfsson and Van Alstyne report that their model indicates, in a direct echo of Schelling's findings, that "other factors being equal, all that is required to reduce integration in most cases is that preferred interactions are more focused than existing interactions." If, in other words, we have even a small inclination to prefer like-minded views and people—to be more "focused" rather than more inclusive—we will tend to create ever more polarized communities online.

We see considerable evidence of such schisms today, particularly in the so-called blogosphere. Political blogs have divided into two clearly defined and increasingly polarized camps: the liberals and the conservatives. In 2005, two researchers, Lada Adamic, of Hewlett–Packard Labs, and Natalie Glance, of Infoseek Applied Research, published the results of an extensive study of political blogs, which they titled "Divided They Blog." They looked at the patterns of linking among the forty most popular political blogs during the two months leading up to the 2004 US presidential election, and they also examined the activity of a much broader set of political blogs—more than 1,000 in all—on one day during that period. They discovered a sharp and "unmistakable" division between the conservative and liberal camps. "In fact," they wrote, "91% of the links originating within either the conservative or liberal communit[y] stay within that community." In addition, the two groups "have different lists of favorite news sources, people, and topics to discuss," with only occasional overlaps.

Another study of the political blogosphere, by Matthew Hindman, a political scientist at Arizona State University, found a similar polarization. Rather than examining the links contained in the blogs, Hindman looked at the actual traffic flows between them. He found that the vast majority of readers tend to stay within the bounds of either the liberal or the conservative sphere. Liberals listen almost exclusively to other liberals, and conservatives listen almost exclusively to other conservatives. "Only a handful of sites," he reports, "share traffic with those on the opposite end of the political spectrum," and the small amount of interaction that does take place between the sides is dominated by what Hindman terms "name calling." His conclusion: "There's not a whole lot of great news for democratic theory here."

DURING THE SUMMER of 2005, a group of researchers assembled sixty-three Coloradans to discuss three controversial issues: same-sex marriage, affirmative action, and global warming. About half of the participants were conservatives from Colorado Springs, while the other half were liberals living in Boulder. After the participants completed, in private, questionnaires about their personal views on the three topics, they were split into ten groups—five conservative and five liberal. Each group then spent some time discussing the issues, with the goal of reaching a consensus on each one. After the discussion, the participants again filled out questionnaires.

The results of the study were striking. In every case, the deliberations among like-minded people produced what the researchers call "ideological amplification." People's views became more extreme and more entrenched:

First, the groups from Boulder became even more liberal on all three issues; the groups from Colorado Springs became even more conservative. Deliberation thus increased extremism. Second, every group showed increased consensus, and decreased diversity, in the attitudes of [its] members. . . . Third, deliberation sharply increased the differences between the views of the largely liberal citizens of Boulder and the largely conservative citizens of Colorado Springs. Before deliberation began, there was considerable overlap between many individuals in the two different cities. After deliberation, the overlap was much smaller.

The study revealed a fact about human nature and group dynamics that psychologists have long recognized: the more that people converse or otherwise share information with other people who hold similar views, the more extreme their views become. As University of Chicago law professor Cass Sunstein, one of the organizers of the Colorado experiment, explains in his book *Infotopia*, "When like-minded people cluster, they often aggravate their biases, spreading falsehoods." They "end up in a more extreme position in line with their tendencies before deliberation began." This phenomenon, which Sunstein reports has been documented "in hundreds of studies in over a dozen countries," may in the worst cases plant "the roots of extremism and even fanaticism and terrorism."

Given how easy it is to find like-minded people and sympathetic ideas on the Internet and given our innate tendency to form homogeneous groups, we can see that "ideological amplification" is likely to be pervasive online. Here again, as Brynjolfsson and Van Alstyne note in their article, filtering and personalization technologies are likely to magnify the effect. "Individuals empowered to screen out material that does not conform to their existing preferences may form virtual cliques, insulate themselves from opposing points of view, and reinforce their biases," they write. "Indulging these preferences can have the perverse effect of intensifying and hardening pre-existing biases. . . . The effect is not merely a tendency for members to conform to the group average but a radicalization in which this average moves toward extremes."

In a further perverse twist, the very abundance of information available on the Internet may serve not to temper extremism but to amplify it further. As the Colorado study showed, whenever people find additional information that supports their existing views, they become more convinced that those views are right—and that people who hold different opinions are wrong. Each extra piece of confirming information heightens their confidence in the rectitude of their opinion and, as their confidence increases, their views tend also to become more extreme. They become single-minded.

Not only will the Internet tend to divide people with different views, in other words, it will also tend to magnify the differences. As Brynjolfsson and Van Alstyne suggest, this could in the long run pose a threat to the spirit of compromise and the practice of consensus-building that are at the heart of democratic government. "Internet users can seek out interactions with like-minded individuals who have similar values and thus become less likely to trust important decisions to people whose values differ from their own," they conclude. Although they stress that it's too early to know exactly how all of these forces will play out, they warn that "balkanization and the loss of shared experiences and values may be harmful to the structure of democratic societies."

THE INTERNET TURNS everything, from news-gathering to community-building, into a series of tiny transactions—expressed mainly through clicks on links—that are simple in isolation yet extraordinarily complicated in the aggregate. Each of us may make hundreds or even thousands of clicks a day, some deliberately, some impulsively, and with each one we are constructing our identity, shaping our influences, and creating our communities. As we spend more time and do more things online, our combined clicks will shape our economy, our culture, and our society.

We're still a long way from knowing where our clicks will lead us. But it's clear that two of the hopes most dear to the Internet optimists—that the Web will create a more bountiful culture and that it will promote greater harmony and understanding—should be treated with skepticism. Cultural impoverishment and social fragmentation seem equally likely outcomes.

CHAPTER 9

Fighting the Net

T OWARD THE END of 2006, British troops stationed in Basra, Iraq, found themselves coming under increasingly accurate mortar fire from insurgent fighters hidden in and around the city. One soldier was killed in the attacks and several others were wounded. During the second week of the new year, the British Army staged a series of raids on the homes and hideouts of suspected insurgents in hopes of curtailing the shelling. When they broke into some of the buildings, they were surprised to discover pages of printouts from the Google Earth mapping service. The printouts showed British positions in enough detail that individual tents and even latrines could be identified. One pictured the headquarters of the 1,000-member Staffordshire Regiment, and written on the back were the camp's longitude and latitude. Army intelligence officers concluded that the insurgents were using the images to aim their artillery.

The discovery confirmed what experts in military technology have long suspected: that terrorists and guerrilla fighters can derive valuable intelligence from Google Earth and other Internet mapping tools. When combined with geopositioning data from common GPS devices, the images can be used to target bombs and assaults with great accuracy. They provide a simple but effective alternative to the high-tech guidance systems employed by modern armies. In a 2005 interview, Brian Collins, a vice president of the British Computer Society, said that "websites like Google Earth give these people the possibility of levelling the playing field a bit. If you can locate a target on the image it will give you very accurate co-ordinates and a terrorist will know exactly where to aim a missile. If you also have a GPS then you know exactly where you are and you can sit there with your PC and look at these very high resolution satellite images and you will know where to fire your missile from and what to fire it at."

The troops who had been the targets of the mortar fire were outraged by the news that their adversaries were using maps and images printed from a public Internet site to aim their shells. They told a reporter from the *Daily Telegraph* that they might sue Google if they were wounded in future attacks. The enemy fighters, bemoaned one soldier, "now have the maps and know exactly where we eat, sleep and go to the toilet."

The soldiers' anger is understandable, but it's misplaced. Google Earth is used by millions of people for entirely benign purposes. Real estate agents use it to display the locations of homes. Teachers use it for geography lessons. Television correspondents use it to illustrate news stories. What it does is no different from what hundreds of other Internet services do: it pulls together publicly available information—photographs from airplanes and satellites, in this case—into a convenient form. The Google service can, as a company spokesman readily admits, be used for "good and bad," but so can any tool. Technology is amoral, and inventions are routinely deployed in ways their creators neither intend nor sanction. In the early years of electrification, electric-shock transmitters developed by the meatpacking industry to kill livestock were appropriated by police forces and spy agencies as tools for torturing people during interrogations. To hold inventors liable for the misuse of their inventions is to indict progress itself.

That's cold comfort if you're facing mortar fire, of course. And the predicament of the British soldiers underscores an important fact that will become more salient in the years ahead: the World Wide Computer is especially prone to misuse and abuse. As a general purpose technology, it puts into the hands of bad guys the same infinitely diverse array of applications it provides to good guys. Computer networks in general and the Web in particular have always been plagued by conmen, criminals, and vandals, who have proven adept at discovering and exploiting vulnerabilities in software, databases, and communication protocols. The scope and scale of the havoc they can wreak only expand as the price of computing and bandwidth falls and as the sharing of data and code becomes more common. The very qualities that make the World Wide Computer so useful to so many—its universality and its openness—make it dangerous as well.

For terrorists, the Internet has been a godsend. In addition to providing easy access to maps, photographs, descriptions of weaponry, and other valuable tactical information, it serves as an all-purpose communications network, surveillance medium, propaganda channel, and recruiting tool, freely available almost anywhere in the world. Researchers at the Dark Web Portal, a University of Arizona project that monitors the online activities of terrorist organizations, were able to uncover more than 325 covert terrorist sites on the Internet in 2006. The sites held some 275 gigabytes of data, including 706 videos (seventy were of beheadings, and twenty-two showed suicide attacks) as well as audio messages and images of prospective and actual attack targets, from buildings to vehicles to pedestrians. The Net provides a ready-made military infrastructure ideally suited to the needs of a widely dispersed, ad hoc, clandestine force.

The US military is well aware of the threat. In October 2003, the Department of Defense and the Joint Chiefs of Staff prepared a secret report, called the Information Operations Roadmap, which declared as one of its main thrusts, "We must fight the net." The Roadmap, then-Defense Secretary Donald Rumsfeld wrote in a foreword, "provides the Department with a plan to advance the goal of information operations as a core military competency. [It] stands as another example of the Department's commitment to transform our military capabilities to keep pace with emerging threats and to exploit new opportunities afforded by innovation and rapidly developing information technologies."

The report, which the government released to the news media in a heavily redacted form in 2005, makes for fascinating if alarming reading. It describes how the US military "is building an information-centric force," with the goal of "dominating the information spectrum." Computer networks "are increasingly the operational center of gravity," the authors write, and "are vulnerable now, and barring significant attention, will become increasingly more vulnerable." The military must "be fully prepared to ensure critical warfighting functionality" online. In particular, the government needs to launch a concerted effort to develop a national policy "on the use of cyberspace for offensive attacks," including a legal review to "determine what level of data or operating system manipulation constitutes an attack" and "which actions can be appropriately taken in self-defense."

The Internet is a battlefield, but it's a battlefield unlike any other. Its boundaries and terrain are in constant flux, sophisticated new weapons can be built and deployed on it by amateur software coders using cheap PCs, and its commercial and social uses are inextricably entwined with its military ones. It neutralizes many of the advantages traditionally held by large armies with cutting-edge physical weapons. "Networks are growing faster than we can defend them," admit the Roadmap's authors. "Attack sophistication is increasing." In one particularly revealing passage, they describe how the Net complicates psychological operations such as the dissemination of propaganda: "Information intended for foreign audiences, including public diplomacy and Psyops, is increasingly consumed by our domestic audience." On the Internet battlefield, neither information nor misinformation stays put. There's no such thing as a local fight.

It's no surprise that a document like the Information Operations Roadmap exists. There's every reason to believe that information networks will be scenes of combat of one sort or another in the future, and a national defense strategy needs to take such possibilities into account. Still, it's disconcerting to read military planners calmly laying out a doomsday scenario in which American forces act to "disrupt or destroy the full spectrum of globally emerging communications systems, sensors, and weapons systems dependent on the electromagnetic spectrum." If the Internet levels the battlefield to the benefit of America's enemies, the only recourse, it seems, may be to destroy the battlefield.

But a high-tech military assault is hardly the only threat facing the Internet. There are many others, from the criminal to the political to the technical. Given its centrality to the world's economy, the Net is a surprisingly insecure infrastructure.

IN A SPEECH at the 2004 World Economic Forum in Davos, Switzerland, a confident Bill Gates declared war on one of the oldest and most intractable scourges of the Internet: junk email. He assured the gathering of business and political dignitaries that "spam will be solved" by 2006. For a brief time, it looked like he might be right.

Sophisticated new filters put a dent in the flow of spam, substantially increasing spammers' costs and threatening to put them out of business. But the spammers fought back. They tapped into the power of the World Wide Computer to launch a new wave of fraudulent emails that made what came before look like child's play. Far from being solved, spam traffic hit new heights. By the end of 2006, an estimated 94 percent of all emails sent over the Internet were spam, up from about 38 percent when Gates made his prediction. One spam-tracking firm reported that on any given day as many as 85 billion spam messages were being sent. Although most of those messages were trapped by filters, enough reached their targets to make the spam business more lucrative than ever.

Why have spammers been able to thrive despite the concerted efforts of the multi-trillion-dollar computing and communications industry to stop them? It's because they, like the rest of us, are able to program the World Wide Computer to do their bidding. Their most potent weapon has become the botnet. Short for "robot network," a botnet is the evil twin of the CERN Grid. It's a large group of privately owned PCs that can be controlled centrally—though the control in this case is wielded not by a research laboratory but by a criminal entity. A botnet is created through the distribution of a virus over the Internet. When the virus finds its way into a PC, through an email attachment or a downloaded file, it installs a small bit of code that allows the machine to be manipulated by instructions from a distant computer. A botnet can include thousands or even millions of "zombie PCs," all acting in concert as a single system—without their owners knowing anything's amiss.

A spammer can use a botnet to send out millions or even billions of messages simultaneously, and because they're funneled through the email programs of ordinary citizens, they often slip past spam filters. Since they commandeer the bandwidth of the PC owners'

Internet accounts, moreover, botnets dramatically reduce the spammers' costs. In effect, the marginal cost of sending out a message falls to zero, making it economical to pump out unlimited quantities of junk. It's believed that between 10 and 25 percent of all computers on the Internet today are infected with botnet viruses and that the zombie networks are responsible for at least 80 percent of all spam.

But while botnets are a nuisance as dispensers of unwanted emails about penny stocks and bootleg pharmaceuticals, they can be put to far darker purposes. A botnet virus can, for instance, scour a PC's hard drive and monitor its user's keystrokes, gathering sensitive personal information and sending it to back over the Internet to its master. The extent of the criminal threat became clear in 2006 when a large file created by a botnet was intercepted by a computer security professional. The file was found to contain private financial data, including credit card numbers and passwords for bank and brokerage accounts, which had been harvested from nearly 800 infected PCs over the course of a month. Noting that a quarter of a million additional PCs are infected with botnet viruses every day, one security-company executive told the *New York Times*, "We are losing this war badly."

Beyond their moneymaking potential, botnets can also be used to sow destruction on the Internet itself. A botnet's master can instruct his army of robot computers to inundate a commercial or governmental site with information requests in a so-called distributed denial of service, or DDoS, attack. Unable to handle the spike in traffic, the site's server crashes, sometimes bringing down an entire data center or even a large subnetwork with it. On the afternoon of May 2, 2006, the American blog-publishing firm Six Apart found itself the victim of a large-scale DDoS assault by a particularly aggressive botnet. Within minutes, the company's servers had crashed, causing the blogs of 10 million of its customers to vanish from the Internet. The attack, Six Apart soon discovered, was aimed not at itself but rather at one of its customers, an Israeli firm named Blue Security. The company sold a service for combating spam, earning it the enmity of the Net's outlaws. The botnet assault went on for days, causing widespread damage to many other companies and sites, until, on May 17, Blue Security surrendered. "We cannot take the responsibility for an ever-escalating cyber war through our continued operations," the company said in a statement. It closed down its business that day and its CEO went into hiding.

Such assaults can be used for political as well as commercial purposes. In the spring of 2007, one or more botnets launched a series of coordinated attacks against Web sites operated by the government of Estonia, rendering some of them inaccessible and slowing traffic to others. Although the source of the assault was never uncovered, it appears to have been launched in retaliation for Estonia's decision to remove a Soviet-era war memorial from the center of Tallinn, its capital. Extending over two weeks, the action involved approximately 130 separate DDoS attacks, some of which lasted more than ten hours. Officials of NATO and the United States, fearful that the assault might serve as a template for future political attacks, sent specialists in online warfare to Estonia to analyze what transpired.

Just before the Estonian action, Bill Gates again spoke at the World Economic Forum. He made no mention, however, of spam and botnets or of the sunny forecast he had delivered three years earlier. But the topic did come up during a panel discussion featuring Vinton Cerf, a former Defense Department computer scientist who played a key role in designing the Internet and who now serves as Google's Chief Internet Evangelist. Warning that more than 100 million computers may now be zombies, Cerf called the spread of botnets a "pandemic." Another panelist, the technology writer John Markoff, agreed. "It's as bad as you can imagine," he told the audience. "It puts the whole Internet at risk."

CLEARLY, THE INTERNET is not just a battlefield. It's also a target—and an extraordinarily high-value one at that. Its strategic importance extends far beyond its potential military role. The Net is rapidly becoming the country's and the world's dominant commercial infrastructure, connecting global supply chains, processing myriad transactions, providing a rich channel for marketing and advertising, and serving as a repository of business and financial data. For the modern information economy, the computing grid is the railway system, the highway network, the electric grid, and the telephone system rolled into one. And, as the Department of Defense's Roadmap makes clear, it's poorly protected. If a remote-controlled botnet can hold a group of companies hostage for days on end, ultimately putting one of them out of business, it's not hard to imagine what a concerted attack by an unfriendly government, a criminal syndicate, or a terrorist organization might do. *Wired* reports that "some 20 nations have computer attack programs under way" and that "one botnet recently detected by Arbor Networks was controlled via chat channels called 'jihad' and 'allah-akbar.'"

In a 2005 letter to President Bush, the members of a distinguished White House advisory committee on information technology delivered a blunt warning. "The IT infrastructure is highly vulnerable to premeditated attacks with potentially catastrophic effects," they wrote. "Thus, it is a prime target for cyber terrorism as well as criminal acts." In subsequent testimony before Congress, one of the committee members, computer scientist and entrepreneur Tom Leighton, put it even more starkly: "Today, virtually every sector of the nation's infrastructure—including communications, utilities, finance, transportation, law enforcement, and defense—is critically reliant on networked IT systems, and these systems have very little, if any, defense against cyber attack. All elements of the nation's infrastructure are insecure if IT is insecure, and, today, our IT is insecure."

Fanatical terrorists and shadowy criminals are not the only threat facing the World Wide Computer. There are more prosaic risks as well, from electricity shortages to natural disasters to technological failures. On December 6, 2006, a group of executives from leading technology companies, including IBM, Google, Silicon Graphics, Cisco, and Hewlett–Packard, sat down in a Silicon Valley conference room for a four-hour meeting with representatives from the US Department of Energy. The discussion centered on the exploding demand for electricity by data centers and the very real possibility that the country's electric system might not be able to keep up with it. The rise of a new national infrastructure threatened to overwhelm an old one. "I think we may be at the beginning of a potential energy crisis for the IT sector," said a Silicon Graphics executive. "It's clearly coming." Google's representative said that unless changes were made in the supply of power, shortages were likely in the next five to ten years.

Andrew Karsner, an assistant secretary of energy, agreed. Calling computer systems "an absolute juggernaut" in energy consumption, he argued that industry and the government shared a "moral obligation" to ensure the country's energy security. "What happens to national productivity when Google goes down for seventy-two hours?" he asked the gathered executives.

Little did Karsner know that just a couple of weeks later, on December 26, he would get a preview of his nightmare scenario—though it was a natural disaster, not a power outage, that did the damage. A large earthquake off the coast of Taiwan severed the main communication cables connecting eastern Asia with the rest of the world. Millions of people found themselves cut off from the Internet, and major business centers like Hong Kong, Seoul, Taipei, and Singapore struggled to keep their financial markets and other commercial services in operation. Hong Kong's airport, a major hub for the continent, was paralyzed as one of its main computer systems failed. Internet service didn't return to normal for weeks. "I haven't experienced anything like this before," said one Chinese executive. "We've become too dependent on these optic fibers—a few of them get damaged, and everything collapses."

The Internet has always had flaws and faced threats—its imminent demise has been predicted by many pundits over the years—but it has proven extraordinarily resilient up to now. As Cerf said at Davos, "the Net is still working, which is amazing." He and the other computer scientists and military planners who designed its underlying structure deserve much of the credit for its robustness. A great deal of redundancy and flexibility has been built into the physical network as well as the communication protocols that shepherd data through it. Traffic can be easily rerouted around bottlenecks or other trouble spots.

But even if we should take predictions of a digital meltdown with a grain of salt, it would be foolish to ignore them, particularly given the stakes involved. If the World Wide Computer crashes, it will, as Karsner implied, take much of the economy down with it. And although the Internet's design has served it well in the past, cracks have begun to appear. Some of the leading experts on computer networking worry that the Net, as currently constructed, may be nearing the end of its useful life. The combined strains of exploding traffic, security breaches, and years of ad hoc technical patches may be too much for it to handle. In a 2005 *Technology Review* article called "The Internet Is Broken," MIT professor David Clark, who served as the Internet's chief protocol architect during most of the 1980s, said, "We might just be at the point where the utility of the Internet stalls—and perhaps turns downward." Princeton computer scientist Larry Peterson concurred, calling today's Net "an increasingly complex and brittle system." Noting that he had recently been called to brief government officials in Washington, he added, with academic understatement, "There is recognition that some of these problems are potentially quite serious."

THE INTERNET IS unlike any commercial or social infrastructure we've ever seen. In contrast to railroads and highways, electric grids and telephone networks, the Net is not a fixed physical system built inside a country's borders and hence under its government's direct control. Many of its most important components—databases, software code, computer processors—are portable. They can be moved around easily, often with just a few clicks of a mouse. Many major American corporations, including General Electric, American Express, Verizon, IBM, and General Motors, have already shifted large chunks of their computing operations to countries like India and China, sometimes putting critical assets and processes under the management of foreign companies and workers. The international transfer of information technology continues to accelerate. The trade journal *CIO Update* reported in 2006 that "small companies are beginning to offshore IT infrastructure (one application at a time), and large multinational companies are offshoring entire data centers."

As computing becomes more of a utility, corporations and even governments will begin to give up not only the management but the ownership of their computing assets, and many of those assets will naturally move to where they can be maintained and operated at the lowest cost. It's not hard to imagine that much of a country's commercial infrastructure could end up scattered throughout the world, under foreign jurisdiction. That raises new and difficult questions about national security and even national sovereignty. Are countries comfortable with the idea of giving up direct control over the machines and software that their economies run on? Do they trust foreign governments, some of which may be unstable or even unfriendly, to safeguard sensitive data about the operations of their companies and the lives of their citizens?

Politicians are only now beginning to wrestle with such questions. In June 2007, for example, France's defense ministry banned the use of BlackBerrys by top government officials. Messages sent with the popular devices are routinely routed through servers in the United States and Britain, and French intelligence officers feared that secret governmental and economic communications could easily be monitored by the US National Security Agency and other foreign groups. "The risks of interception are real," one intelligence minister told *Le Monde*.

While US lawmakers are coming to appreciate the military, criminal, and technical threats to the Internet's stability, they still seem oblivious to issues of political control. There are several reasons for the lack of concern. For one thing, few elected officials have a strong understanding of how the Internet works. They don't know the code. For another, most of the components of the new digital infrastructure have been built by and remain in the hands of thousands of private companies and universities. They lie outside the purview of government bureaucrats and legislators. Until recently, the Internet hasn't even looked like a national infrastructure—it seemed to be just a loose array of computer systems housed mainly in private data centers and connected by cables hidden in the ground.

Finally, those aspects of the Internet that have required regulation, such as the setting of standards and the choosing of protocols, have remained under de facto American control. The Net had its origins within the US military and academic establishment; most of the major computing and networking firms, from IBM and Microsoft to Cisco and Google, have been headquartered here; and the Net's chief oversight bodies have tended to be dominated by American interests. The powerful Internet Corporation for Assigned Names and Numbers (ICANN), which oversees the assignment of domain names and addresses—the valuable "real estate" of the Net—has long operated under the auspices of the US Department of Commerce, even though it's a putatively international body. And most of the Net's thirteen "root servers," the computers that ultimately control the routing of all traffic, remain in the hands of US government agencies and corporations.* Because the network has always felt like "our" Internet, Americans and their lawmakers have been sanguine about its operation and control.

There are signs, however, that American hegemony will not remain uncontested. Authoritarian countries like China, Saudi

* Every computer on the Internet has a unique Internet Protocol (IP) address, which takes the form of four sets of digits (for example, 123.123.123.123). Because people find it hard to remember long strings of numbers (and because IP addresses are subject to change), the computers that run Internet sites are also given familiar, stable domain names (for example, www.whitehouse.gov). Scattered throughout the Net are thousands of "name servers" that translate the domain names into IP addresses. The thirteen root servers stand at the top of the hierarchy of name servers, as they contain information about the location of addresses for the Net's top-level domains, such as .com, .org, and .uk. If you enter "www.whitehouse.gov" into your Web browser, for instance, the root-server database will direct your computer to another name server that stores the IP addresses of sites within the .gov domain. ICANN oversees the assignment of IP addresses and domain names and the maintenance of the root-server database. ICANN's decisions have great economic value, as well as political and cultural import, because they determine how the Net's property is carved up, in terms of ownership and control, and how traffic flows throughout the network.

Arabia, and Iran, which have always been uncomfortable with the Internet's openness, have grown more aggressive in demanding that individual states be given greater sovereignty over the Net. Other countries are also calling for changes in the governance structure, going so far as to suggest that America's authority over the assignment of site addresses and names and the routing of traffic represents a form of "neo-colonialism." Brazil, frustrated with the lack of control over online pornography, threatened to carve out its own regional Internet if the current system isn't changed. In September 2005, the renegades' cause received a big boost when the European Union surprised the United States by coming out in favor of greater international oversight of the Net, calling for "the establishment of an arbitration and dispute resolution mechanism based on international law." The growing tensions dominated discussions at the UN-sponsored World Summit on the Information Society held in Tunisia in November 2005. Unable to reach any concrete agreement, the delegates issued a vague statement stressing that "any Internet governance approach should be inclusive and responsive and should continue to promote an enabling environment for innovation, competition and investment."

As the importance of the Internet as a shared global infrastructure grows, decisions about its governance as well as its structure and protocols will take on greater weight. The new computing grid may span the world, but as the dominant medium for commerce, communication, and even culture it has profound national and regional implications. Very different conceptions of how the grid should operate are emerging, reflecting the economic, political, and social interests of different countries and regions. Soon, governments will be forced to start picking sides. They will have to choose, as Jack Goldsmith and Tim Wu write in *Who Controls the Internet?*, among systems of governance "ranging from the United

States's relatively free and open model to China's model of political control." Seemingly arcane technical standards, originally intended to create unity, will become the new terms of ideological debate and geopolitical struggle. The result, argue Goldsmith and Wu, will be "a technological version of the cold war, with each side pushing its own vision of the Internet's future."

Whether we're on the verge of Cold War 2.0 or not, the years ahead promise to be perilous ones as states and citizens struggle to come to grips with the manifold ramifications of the universal computing grid. As the Venezuelan scholar Carlota Perez has shown, governments tend to be very slow to respond to technological revolutions. Even as entrepreneurs and financiers, not to mention criminals and other bad actors, rush to exploit commercial and political disruptions, politicians, judges, and bureaucrats remain locked in the past, pursuing old policies and relying on outdated legal and regulatory schemes. The inertia magnifies the social and economic uncertainty and upheaval. In the worst cases, it lasts for decades, exacting, as Perez puts it, "a very high cost in human suffering."

CHAPTER 10

A Spider's Web

WHO IS 4417749?

That was the question two *New York Times* reporters, Michael Barbaro and Tom Zeller Jr., set out to answer on August 7, 2006. In late July, AOL had released through its Web site a report detailing the keywords entered into its search engine by 657,000 of its subscribers over a three-month period earlier in the year. The company, a unit of media conglomerate Time Warner, thought it was performing a public service by releasing the search logs. It knew the information would be valuable to academic and corporate researchers studying the behavior of Web surfers or trying to invent new search technologies. To protect subscribers' privacy, AOL had carefully "anonymized" the data, replacing people's names with numbers and stripping out other identifying information. Said one Stanford computer science professor, "Having the AOL data available is a great boon for research."

But others wondered whether the data was really as anonymous as it seemed. Could the identities of the subscribers be inferred simply by examining what they searched for? Barbaro and Zeller, along with their editor, David Gallagher, decided to find out. They took a close look at one set of keywords, those entered by an AOL subscriber known only as "4417749." The terms were a mishmash, ranging from "swing sets" to "single dances in Atlanta" to "dog who urinates on everything" to "school supplies for Iraq children." They formed what the reporters called "a catalog of intentions, curiosity, anxieties and quotidian questions." But there were enough clues in that catalog for Barbaro, Zeller, and Gallagher to track down the name, address, and phone number of the searcher. The search took only "a couple of hours," according to Gallagher. Number 4417749 turned out to be Thelma Arnold, a sixty-two-year-old widow living in Lilburn, Georgia. On August 9, Arnold woke up to find her name and picture on the front page of the national edition of the *Times*.

She was shocked to discover that her searches had been monitored by AOL, each keyword meticulously collected and connected to her account. "My goodness, it's my whole personal life," she told the reporters. "I had no idea somebody was looking over my shoulder." But however embarrassing she found the disclosure, Arnold had some cause to be relieved. The terms she had searched for were innocuous. Other subscribers had divulged much more intimate information about themselves. Subscriber 11574916 searched for "cocaine in urine" and "florida dui laws." Subscriber 1515830 searched for "how to tell your family you're a victim of incest" and "can you adopt after a suicide attempt." Subscriber 59920 searched for "what a neck looks like after its been strangled" and "rope to use to hog tie someone." Along with the quotidian came the peculiar and the perverse.

Like Thelma Arnold, most of us assume that we're anonymous when we go about our business online. We treat the Internet not just as a shopping mall and a library but as a personal diary and even a confessional. Through the sites we visit and the searches we make, we disclose details not only about our jobs, hobbies, families, politics, and health but also about our secrets, fantasies, obsessions, peccadilloes, and even, in the most extreme cases, our crimes. But our sense of anonymity is largely an illusion. Detailed information about everything we do online is routinely gathered, stored in corporate or governmental databases, and connected to our real identities, either explicitly through our user names, our credit card numbers, and the IP addresses automatically assigned to our computers or implicitly through our searching and surfing histories. A famous 1993 *New Yorker* cartoon bore the caption "On the Internet, nobody knows you're a dog." In reality, not only is it known that you're a dog, but it's probably known what breed you are, your age, where you live, and the kind of treat you prefer.

Linking sensitive information to people's names doesn't require a team of *New York Times* reporters sifting through search logs and phone books. Nor does it require the inadvertent or intentional disclosure of data. As online databases proliferate and as analytical technologies advance, it becomes ever easier to use the World Wide Computer to "mine" personal information. A few months before AOL released its search logs, the writer Tom Owad provided a chilling lesson in just how easy it has become to extract private data from the Internet. Owad, who publishes a Web site for Macintosh users, wrote a simple piece of software—a "script"—to download the wish lists posted by Amazon.com customers. Millions of people maintain such lists on the online retailer's site, using them to catalog products that they plan to purchase in the future or that they'd like to receive as gifts. These lists can be searched by anyone, and they usually include the name of the list's owner and the city and state in which he lives.

Using two five-year-old PCs and a standard home Internet connection, Owad was able to download more than a quarter of a million wish lists over the course of a day. "I now had documents describing the reading preferences of 260,000 US citizens," he later wrote on his site. Encoded into each list was the owner's unique Amazon customer identification number, allowing Owad to easily sort the lists by individual. He could then search the resulting database for book titles as well as various other keywords. He performed searches for several controversial or politically sensitive books and authors, from Kurt Vonnegut's *Slaughterhouse-Five* to the Koran, from the right-wing pundit Rush Limbaugh to his left-wing counterpart Michael Moore. Knowing the names and home cities of the list owners, he was then able to use Yahoo People Search to identify addresses and phone numbers for many of them. He took one final step and used Google Maps to plot their street addresses. He ended up with maps of the United States showing the precise locations of people interested in particular books and ideas. He posted on his site, for instance, a map of the homes of Amazon customers who had expressed an interest in George Orwell's *1984*. He could just as easily have published a map showing the residences of people interested in books about growing marijuana or giving up a child for adoption. "It used to be," Owad concluded, "you had to get a warrant to monitor a person or a group of people. Today, it is increasingly easy to monitor ideas. And then track them back to people."

Owad spent a fair amount of time organizing and conducting his information-gathering experiment. He had to write a custom code to download the data, and he had to manually perform his database searches. But what Owad did by hand can increasingly be performed automatically, with data-mining algorithms that draw information from many different sites simultaneously. One of the essential characteristics of the computing grid is the interconnection of diverse stores of information. The "openness" of databases is what gives the World Wide Computer much of its power. But it also makes it easy to discover hidden relationships among far-flung bits of data. Analyzing those relationships can unlock a surprisingly large trove of confidential information about Web users.

At a 2006 conference of computer scientists, held in Seattle during the same week that Thelma Arnold's identity was revealed in the press, five scholars from the University of Minnesota presented a paper titled "You Are What You Say: Privacy Risks of Public Mentions." They described how software programs can be used to make connections among online databases. By uncovering overlaps in the data, the programs can often create detailed personal profiles of individuals—even when they submit information anonymously. The software is based on a simple and obvious principle: people tend to express their interests and discuss their opinions in many different places on the Internet. They may, for instance, buy an album at the iTunes Music Store, include that album on their playlist at Last.fm, rate it at the Rate Your Music site, and mention it in a comment on a music blog. Or they may edit a Wikipedia entry on their favorite actor, write a review of a new biography of him at Amazon, become his "friend" at MySpace, and tag pictures of him at Flickr. Sophisticated algorithms can identify such correspondences and use them to identify individuals with extraordinary precision, the Minnesota researchers discovered. In analyzing just two databases—one drawn from a movie ratings site, the other from a movie discussion forum—they found that an algorithm could successfully identify 60 percent of the people who mentioned eight or more films.

"In today's data-rich networked world," they explain in their paper, "people express many aspects of their lives online. It is common to segregate different aspects in different places: you might write opinionated rants about movies in your blog under a pseudonym while participating in a forum or Web site for scholarly discussion of medical ethics under your real name. However, it may be possible to link these separate identities" using data-mining algorithms. Such an automated process of identification, they say, "creates serious privacy risks for users." Even if people don't divulge their real identities anywhere, their names can often be easily discovered if they disclose a very small number of identifying characteristics. The authors note, for example, that the vast majority of Americans can be identified by name and address using only their zip code, birthday, and gender—three pieces of information that people routinely divulge when they register for a user name at a Web site.

"You have zero privacy," Scott McNealy, the former chief executive of Sun Microsystems, remarked back in 1999. "Get over it." The idea that the loss of privacy is the price we have to pay for the handiness of the Internet is a common one, and there's some truth to it. But few of us are aware of the extent to which we've disclosed details about our identities and lives or the way those details can be mined from search logs or other databases and linked back to us. And whether or not we're comfortable with the possible compromise of our privacy, that's far from the only or even the most disturbing threat posed by today's Internet. As mathematicians and computer scientists continue to refine data-mining algorithms, they are uncovering new ways to predict how people will react when they're presented with information or other stimuli online. They're learning not just how to identify us but how to manipulate us, and their discoveries are being put to practical use by companies and governments, not to mention con artists and crooks.

It's natural to think of the Internet as a technology of emancipation. It gives us unprecedented freedom to express ourselves, to share our ideas and passions, to find and collaborate with soul mates, and to discover information on almost any topic imaginable. For many people, going online feels like a passage into a new and radically different kind of democratic state, one freed of the physical and social demarcations and constraints that can hobble us in the real world. The sense of the Web as personally "empowering," to use the common buzzword, is almost universal, even among those who rue its commercialization or decry the crassness of much of its content. In early 2006, the editors of the Cato Institute's online journal *Cato Unbound* published a special issue on the state of the Net. They reported that the "collection of visionaries" contributing to the issue appeared to be "in unanimous agreement that the Internet is, and will continue to be, a force for liberation." In a July 2007 essay, the media scholar Clay Shirky wrote, "The internet's output is data, but its product is freedom, lots and lots of freedom." David Weinberger, in his book *Small Pieces Loosely Joined,* summed up the Internet's liberation mythology in simple terms: "The Web is a world we've made for one another."

It's a stirring thought, but like most myths it's at best a half-truth and at worst a fantasy. Computer systems in general and the Internet in particular put enormous power into the hands of individuals, but they put even greater power into the hands of companies, governments, and other institutions whose business it is to control individuals. Computer systems are not at their core technologies of emancipation. They are technologies of control. They were designed as tools for monitoring and influencing human behavior, for controlling what people do and how they do it. As we spend more time online, filling databases with the details of our lives and desires, software programs will grow ever more capable of discovering and exploiting subtle patterns in our behavior. The people or organizations using the programs will be able to discern what we want, what motivates us, and how we're likely to react to various stimuli. They will, to use a cliché that happens in this case to be true, know more about us than we know about ourselves.

Even as the World Wide Computer grants us new opportunities and tools for self-expression and self-fulfillment, it is also giving others an unprecedented ability to influence how we think and what we do, to funnel our attention and actions toward their own ends. The technology's ultimate social and personal consequences will be determined in large measure by how the tension between the two sides of its nature—liberating and controlling—comes to be resolved.

ALL LIVING SYSTEMS, from amoebas to nation-states, sustain themselves through the processing of matter, energy, and information. They take in materials from their surroundings, and they use energy to transform those materials into various useful substances, discarding the waste. This continuous turning of inputs into outputs is controlled through the collection, interpretation, and manipulation of information. The process of control itself has two thrusts. It involves measurement—the comparison of the current state of a system to its desired state. And it involves two-way communication—the transmission of instructions and the collection of feedback on results. The processing of information for the purpose of control may result in the release of a hormone into the bloodstream, the expansion of a factory's production capacity, or the launch of a missile from a warship, but it works in essentially the same way in any living system.

When Herman Hollerith created his punch-card tabulator in the 1880s, he wasn't just pursuing his native curiosity as an engineer and an inventor. He was responding to an imbalance between, on the one hand, the technologies for processing matter and energy and, on the other, the technologies for processing information. He was trying to help resolve what James R. Beniger, in *The Control Revolution*, calls a "crisis of control," a crisis that threatened to undermine the stability of markets and bring economic and technological progress to a standstill.

Throughout the first two centuries of the Industrial Revolution, the processing of matter and energy had advanced far more rapidly than the processing of information. The steam engine, used to power ships and trains and industrial machines, allowed factories, transportation carriers, retailers, and other businesses to expand their operations and their markets far beyond what was possible when production and distribution were restricted by the limitations of physical strength. Business owners, who had previously been able to observe their operations in their entirety and control them directly, now had to rely on information from many different sources to manage their companies. But they found that they lacked the means to collect and analyze the information fast enough to make timely decisions. Measurement and communication both began to break down, hamstringing management and impeding the further growth of businesses. As the sociologist Emile Durkheim observed in 1893, "The producer can no longer embrace the market in a glance, nor even in thought. He can no longer see limits, since it is, so to speak, limitless. Accordingly production becomes unbridled and unregulated." Government officials found themselves in a similar predicament, unable to assemble and analyze the information required to regulate commerce. The processing of materials and energy had progressed so rapidly that it had gone, quite literally, out of control.

During the second half of the nineteenth century, a series of technological advances in information processing helped administrators, in both business and government, begin to reimpose control over commerce and society, bringing order to chaos and opening the way for even larger organizations. The construction of the telegraph system, begun by Samuel F. B. Morse in 1845, allowed information to be communicated instantaneously across great distances. The establishment of time zones in 1883 allowed for more precise scheduling of trains, speeding shipments and reducing accidents. The most important of the new control technologies, however, was bureaucracy—the organization of people into hierarchical information-processing systems. Bureaucracies had been around as long as civilization itself, but as Beniger writes, "bureaucratic administration did not begin to achieve anything approximating its modern form until the late Industrial Revolution." Just as the division of labor in factories provided for the more efficient processing of matter, so the division of labor in government and business offices allowed for the more efficient processing of information.

But bureaucrats alone could not keep up with the flood of data that needed to be processed—the measurement and communication requirements went beyond the capacities of even large groups of human beings. Just like their counterparts on factory floors, information workers needed new tools to do their jobs. That requirement became embarrassingly obvious inside the US Census Bureau at the end of the century. During the 1870s, the federal government, struggling to administer a country and an economy growing rapidly in size and complexity, had demanded that the Bureau greatly expand the scope of its data collection, particularly in the areas of business and transport. The 1870 census had spanned just five subjects; the 1880 round was expanded to cover 215. But the new census turned into a disaster for the government. Even though many professional managers and clerks had been hired by the Bureau, the volume of data overwhelmed their ability to process it. By 1887, the agency found itself in the uncomfortable position of having to begin preparations for the next census even as it was still laboring to tabulate the results of the last one. It was in that context that Hollerith, who had worked on the 1880 tally, rushed to invent his information-processing machine. He judged, correctly, that it would prove invaluable not only to the Census Bureau but to large companies everywhere.

The arrival of Hollerith's tabulator was a seminal event in a new revolution—a "Control Revolution," as Beniger terms it—that followed and was made necessary and inevitable by the Industrial Revolution. Through the Control Revolution, the technologies for processing information finally caught up with the technologies for processing matter and energy, bringing the living system of society back into equilibrium. The entire history of automated data processing, from Hollerith's punch-card system through the mainframe computer and on to the modern computer network, is best understood as part of that ongoing process of reestablishing and maintaining control. "Microprocessor and computer technologies, contrary to currently fashionable opinion, are not new forces only recently unleashed upon an unprepared society," writes Beniger, "but merely the latest installment in the continuing development of the Control Revolution."

It should come as no surprise, then, that most of the major advances in computing and networking, from Hollerith's time to the present, have been spurred not by a desire to liberate the masses but by a need for greater control on the part of commercial and governmental bureaucrats, often ones associated with military operations and national defense. Indeed, the very structure of a bureaucracy is reflected in the functions of a computer. A computer gathers information through its input devices, records information as files in its memory, imposes formal rules and procedures on its users through its programs, and communicates information through its output devices. It is a tool for dispensing instructions, for gathering feedback on how well those instructions are carried out, and for measuring progress toward some specified goal. In using a computer, a person becomes part of the control mechanism. He turns into a component of what the Internet pioneer J. C. R. Licklider, in his seminal 1960 paper "Man–Computer Symbiosis," described as a system integrating man and machine into a single, programmable unit.

But while computer systems played a major role in helping businesses and governments reestablish central control over workers and citizens in the wake of the Industrial Revolution, the other side of their nature—as tools for personal empowerment—has also helped shape modern society, particularly in recent years. By shifting power from institutions to individuals, information-processing machines can dilute and disturb control as well as reinforce it. Such disturbances tend to be short-lived, however. Institutions have proven adept at reestablishing control through the development of ever more powerful information technologies. As Beniger explains, "information processing and flows need themselves to be controlled, so that informational technologies continue to be applied at higher and higher levels of control."

The arrival of the personal computer in the 1980s posed a sudden and unexpected threat to centralized power. It initiated a new, if much more limited, crisis of control. Pioneered by countercultural hackers and hobbyists, the PC was infused from the start with a libertarian ideology. As memorably portrayed in Apple Computer's dramatic "1984" television advertisement, the personal computer was to be a weapon against central control, a tool for destroying the Big Brother-like hegemony of the corporate mainframe and its dominant producer, IBM. Office workers began buying PCs with their own money, bringing them to their offices, and setting them up on their desks. Bypassing corporate systems altogether, PC-empowered employees seized control of the data and programs they used. They gained freedom, but in the process they weakened the ability of bureaucracies to monitor and steer their work. Business executives and the IT managers that served them viewed the flood of PCs into the workplace as "a Biblical plague," in the words of Paul Ceruzzi, the computer historian.

The breakdown of control proved fleeting. The client–server system, which tied the previously autonomous PCs together into a network connected to a central store of corporate information and software, was the means by which the bureaucrats reasserted their control over information and its processing. Together with an expansion in the size and power of IT departments, client–server systems enabled companies to restrict access to data and to limit the use of software to a set of prescribed programs. Ironically, once they were networked into a corporate system, PCs actually allowed companies to monitor, structure, and guide the work of employees more tightly than ever. "Local networking took the 'personal' out of personal computing," explains Ceruzzi. "PC users in the workplace accepted this Faustian bargain. The more computer-savvy among them resisted, but the majority of office workers hardly even noticed how much this represented a shift away from the forces that drove the invention of the personal computer in the first place. The ease with which this transition took place shows that those who believed in truly autonomous, personal computing were perhaps naïve."

The popularization of the Internet, through the World Wide Web and its browser, touched off a similar control crisis. Although the construction of the Internet was spearheaded by the Defense Department, a paragon of centralized power, it was designed, paradoxically, to be a highly dispersed, loosely organized network. Since the overriding goal was to build as reliable a system as possible—one that could withstand the failure of any of its parts—it was given a radically decentralized structure. Every computer, or node, operates autonomously, and communications between computers don't have to pass through any central clearinghouse. The Net's "internal protocols," as New York University professor Alexander Galloway writes, "are the enemy of bureaucracy, of rigid hierarchy, and of centralization." If a corporate computer network was akin to a railroad, with tightly scheduled and monitored traffic, the Internet was more like the highway system, with largely free-flowing and unsupervised traffic.

At work and at home, people found they could use the Web to once again bypass established centers of control, whether corporate bureaucracies, government agencies, retailing empires, or media conglomerates. Seemingly uncontrolled and uncontrollable, the Web was routinely portrayed as a new frontier, a Rousseauian wilderness in which we, as autonomous agents, were free to redefine society on our own terms. "Governments of the Industrial World," proclaimed John Perry Barlow in his 1996 manifesto "A Declaration of the Independence of Cyberspace," "you are not welcome among us. You have no sovereignty where we gather." But, as with the arrival of the PC, it didn't take long for governments and corporations to begin reasserting and even extending their dominion.

The error that Barlow and many others have made is to assume that the Net's decentralized structure is necessarily resistant to social and political control. They've turned a technical characteristic into a metaphor for personal freedom. But, as Galloway explains, the connection of previously untethered computers into a network governed by strict protocols has actually created "a new apparatus of control." Indeed, he writes, "the founding principle of the Net is control, not freedom—control has existed from the beginning." As the disparate pages of the World Wide Web turn into the unified and programmable database of the World Wide Computer, moreover, a powerful new kind of control becomes possible. Programming, after all, is nothing if not a method of control. Even though the Internet still has no center, technically speaking, control can now be wielded, through software code, from anywhere. What's different, in comparison to the physical world, is that acts of control become harder to detect and those wielding control more difficult to discern.

IN EARLY 2000, a Frenchman named Mark Knobel sued Yahoo for selling Nazi memorabilia through its online auction pages. Distributing such goods had long been illegal in France, and Knobel saw no reason why the law shouldn't apply to Internet merchants as it applied to local shopkeepers. "There is this naïve idea that the Internet changes everything," said one of Knobel's lawyers in filing the suit in a Paris court. "It doesn't change everything. It doesn't change the laws in France." But Yahoo's founder Jerry Yang, viewing the suit as a silly and futile attack on the supranational imperium of the Internet, believed that it was Knobel and his lawyers who were the ones afflicted by naïveté. "The French tribunal wants to impose a judgment in an area over which it has no control," he said. "Asking us to filter access to our sites is very naïve."

The suit went forward, nonetheless, and Knobel won. On November 20, 2000, a French judge ruled that Yahoo had broken the law and ordered the company to use its "best effort" to remove Nazi merchandise from any Web page that could be viewed in France. He noted that the company, contrary to its initial claims, was already using software to identify the locations of visitors to its sites in order to serve them customized advertisements. If it could control the ads it showed to people in different countries, he reasoned, it could control the other content they saw as well.

Unbowed, Yahoo announced that it would ignore the decision, claiming that the French courts had no authority in the matter. The judge was not amused. If the company did not comply with his order by February 1, 2001, he said, Yahoo's French assets would be subject to seizure and its executives would be subject to arrest if they set foot in Europe. At that point, Yahoo retreated. A public company with interests around the world, it had little choice but to give in. On January 2, it announced that it would ban from all its sites the sale of products "associated with groups which promote or glorify hatred and violence."

A year later, in the summer of 2002, Jerry Yang had little to say, at least publicly, when Yahoo signed an agreement with the Chinese government that required the company to actively monitor and censor the contents of its sites in China. Yang and other Yahoo executives also remained quiet in 2005 when the company obeyed a demand by Chinese authorities that it reveal the identity of a Chinese citizen who had used a Yahoo email account to send a message about the anniversary of the Tiananmen Square massacre to an organization in the United States. A journalist named Shi Tao was promptly arrested and jailed. "The Yahoo story," write Jack Goldsmith and Tim Wu, "encapsulates the Internet's transformation from a technology that resists territorial law to one that facilitates its enforcement."

Not only are governments in general beginning to partition the online world along old geopolitical lines, but authoritarian regimes are coming to realize that the Internet may not pose as large a threat to their power as they initially feared. While the Net offers people a new medium for discovering information and voicing opinions, it also provides bureaucrats with a powerful new tool for monitoring speech, identifying dissidents, and disseminating propaganda. In a country like China, anyone who assumes that he can act anonymously on the Web opens himself to dangers far beyond embarrassment. In a 2007 speech, China's president, Hu Jintao, spoke glowingly of the Internet's potential for reinforcing the Communist Party's influence over the thoughts of his country's people.

"Strengthening network culture construction and management," he told a group of high-ranking party officials, "will help extend the battlefront of propaganda and ideological work. It is good for increasing the radiant power and infectiousness of socialist spiritual growth."

Democratic governments, as well, have begun scouring Web databases and monitoring Internet traffic for purposes of domestic surveillance. In 2004, federal agencies in the United States were conducting or planning 199 data-mining programs, according to a survey by congressional auditors. In late 2005 and early 2006, press reports indicated that the shadowy National Security Agency had been tapping into the commercial switches that route Internet traffic in a sweeping data-mining operation aimed at uncovering terrorists. As phone calls and other conversations are digitized and routed over the Internet and as geopositioning chips proliferate, the ability of governments of all stripes to monitor their citizens' words and movements will only increase.

Businesses have also found that the Internet, far from weakening their control over employees, actually strengthens their hand. Corporate influence over the lives and thoughts of workers used to be bounded by both space and time. Outside the walls of a company's offices and outside the temporal confines of the workday, people were largely free from the control of their bosses. But one of the consequences of the Net's boundary-breaking is that the workplace and the workday have expanded to fill all space and all time. Today, corporate software and data can be accessed from anywhere over the Internet, and email and instant-messaging traffic continues around the clock. In many companies, the de facto assumption is that employees are always at work, whether they're in their office, at their home, or even on vacation.

The BlackBerry has become the most visible symbol of the expansion of corporate control over people's lives. Connected wirelessly to corporate servers, the ubiquitous gadget forms an invisible tether tying employees to their jobs. For many of today's knowledge workers, turning off the BlackBerry is the last thing they do before going to bed and turning it back on is the first thing they do upon waking. The *Wall Street Journal,* in a 2006 feature titled "BlackBerry Orphans," told the story of a typical BlackBerry-addicted executive whose young children demanded that she not check her email while at home with them in the evening. "To get around their dictates," reported the paper, "the mother hides the gadget in the bathroom, where she makes frequent trips before, during and after dinner." The woman sheepishly told the *Journal* that her children "think I have a small bladder."

The story is as amusing as it is poignant, but it underscores the striking change in the relationship between employers and employees that the Net has already produced. And it reveals another Faustian bargain that employees have struck with computer technology. Many people feel a genuine sense of empowerment when they use their BlackBerry or otherwise connect from afar to their corporate network. They welcome the technology because it "frees" them to work whenever and wherever they want, making them more productive and successful in their jobs. The price they pay, of course, is a loss of autonomy, as their employers gain greater control over their time, their activities, and even their thoughts. "Even though I'm home," another BlackBerry user told the *Journal,* "I'm not necessarily there."

With the data-collection and analytical tools of the World Wide Computer, employers will be able to extend their influence even further. Some companies have already begun to create mathematical models of their workforces, reducing each employee to a set of numbers that can be "optimized" by computer algorithms. IBM, which in recent years has developed sophisticated software for modeling the workings of industrial supply chains, is now creating similar models for managing people. A forty-person group of IBM statisticians and data-mining experts is working "to refocus the supply-chain programs on 50,000 of the consultants in IBM's services division," reports *BusinessWeek*. "That means that instead of modeling machines, furnaces, and schedules, they're building models of their colleagues." The team is drawing information on employees from IBM's many corporate databases, and it's also looking at incorporating data from workers' email messages, online calendars, and mobile phone calls. The company hopes to use the model to assign consultants automatically to projects and to direct their work for optimum efficiency.

Google, too, has launched an experiment in using mathematical modeling for personnel management. During the summer of 2006, it asked its employees to fill out an extensive online survey about themselves, answering some 300 questions on everything from the programming languages they know to the magazines they read to the pets they keep. The company ran the answers through a computer, comparing them to various measures of the employees' skills and achievements in order to create algorithms that might predict performance. In 2007, it began using the algorithms to evaluate all job applicants, who are now also required to fill out a long questionnaire on the Internet. As such modeling techniques progress, they will come to be used much more broadly by companies. "This mathematical modeling of humanity promises to be one of the great undertakings of the 21st century," *BusinessWeek* concludes. "And it doesn't take much imagination to see where that can lead. Managers will operate tools not only to monitor employees' performance but also to follow their movements and drive up productivity."

———

THE MOST far-reaching corporate use of the World Wide Computer as a control technology will not be for optimizing what we do as employees. It will be for optimizing how we act as consumers. Despite the resistance of the Web's early pioneers and pundits, consumerism long ago replaced libertarianism as the prevailing ideology of the online world. Restrictions on the commercial use of the Net collapsed with the launch of the World Wide Web in 1991. The first banner ad—for a Silicon Valley law firm—appeared in 1993, followed the next year by the first spam campaign. In 1995, Netscape tweaked its Navigator browser to support the "cookies" that enable companies to identify and monitor visitors to their sites. By 1996, the dotcom gold rush had begun. More recently, the Web's role as a sales and promotion channel has expanded further. Assisted by Internet marketing consultants, companies large and small have become much more adept at collecting information on customers, analyzing their behavior, and targeting products and promotional messages to them.

The growing sophistication of Web marketing can be seen most clearly in advertising. Rather than being dominated by generic banner ads, online advertising is now tightly tied to search results or other explicit indicators of people's desires and identities. Search engines themselves have become the leading distributors of ads, as the prevailing tools for Web navigation and corporate promotion have merged into a single and extraordinarily profitable service. Google originally resisted the linking of advertisements to search results—its founders argued that "advertising-funded search engines will be inherently biased towards the advertisers and away from the needs of the consumers"—but now it makes billions of dollars through the practice. Search-engine optimization—the science of using advanced statistical techniques to increase the likelihood that a person will visit a site or click on an ad—has become an important corporate function, which Google and other search engines promote by sharing with companies information on how they rank sites and place ads.

In what is perhaps the most remarkable manifestation of the triumph of consumerism on the Web, popular online communities like MySpace encourage their members to become friends with corporations and their products. During 2006, for example, more than 85,000 people "friended" Toyota's Yaris car model at the site, happily entangling themselves in the company's promotional campaign for the recently introduced vehicle. "MySpace can be viewed as one huge platform for 'personal product placement,'" writes Wade Roush in an article in *Technology Review*. He argues that "the large supply of fake 'friends,' together with the cornucopia of ready-made songs, videos, and other marketing materials that can be directly embedded in [users'] profiles, encourages members to define themselves and their relationships almost solely in terms of media and consumption." In recognition of the blurring of the line between customer and marketer online, *Advertising Age* named "the consumer" its 2007 Advertising Agency of the Year.

But the Internet is not just a marketing channel. It's also a marketing laboratory, providing companies with unprecedented insights into the motivations and behavior of shoppers. Businesses have long been skilled at controlling the supply side of their operations, thanks in large part to earlier advances in information technology, but they've struggled when it comes to exerting control over the demand side—over what people buy and where and when they buy it. They haven't been able to influence customers as directly as they've been able to influence employees and suppliers. Advertising and promotion have always been frustratingly imprecise. As the department store magnate John Wanamaker famously said more than a hundred years ago, "Half the money I spend on advertising is wasted. The trouble is, I don't know which half."

The World Wide Computer is beginning to change that. It promises to strengthen companies' control over consumption by providing marketers with the data they need to personalize their pitches
precisely and gauge the effects of those pitches accurately. It opti-
mizes both communication and measurement. In a 2006 interview
with the *Economist*, Rishad Tobaccowala, a top executive with the
international ad agency Publicis, summed up the change in a colorful, and telling, metaphor. He compared traditional advertising to
dropping bombs on cities—a company can't be sure who it hits and
who it misses. But with Internet ads, he said, companies can "make
lots of spearheads and then get people to impale themselves."

From time to time, in response to public or governmental concerns, the leading search engines and other top Internet companies roll out new "privacy safeguards" with great fanfare. But the
moves are rarely more than window-dressing. In the summer of
2007, Google announced it would delete the information it stores on
a person's searches after two years—but only if the person doesn't
perform a Google search or visit any other Google site during that
period, an unlikely scenario given the company's dominance on the
Web. Around the same time, Microsoft issued a press release saying
it would "anonymize" data on searches after eighteen months. As
Thelma Arnold's experience shows, anonymization provides little
real protection in the face of sophisticated data-mining techniques.
In fact, even as Microsoft was making its announcement, the company had a team of Chinese researchers hard at work developing
analytical software for distilling personal demographic information from anonymous online data. The team had already developed
an algorithm able to predict with considerable accuracy a Web
surfer's age and gender based on the sites he or she visits, and it was
working on prediction algorithms for other characteristics such as
occupation, education, and location. In a paper on their work, the

researchers wrote that "the diversity of [a] user's online browsing activities can be exploited to determine an unknown user's demographic attributes."

The ability of businesses to gather and analyze rich data on individual customers comes at the same time that psychologists and economists are making progress in a new discipline called neuromarketing. Neuromarketers use brain scans to uncover the mental triggers that determine what we buy. In a landmark 2007 article, "Neural Predictors of Purchases," published in the journal *Neuron*, a group of scholars from MIT, Stanford, and Carnegie Mellon reported that they could use MRI machines to monitor the brain activity of shoppers as they evaluated products and prices on computer screens. By pinpointing which circuits in the brain "lit up" at different stages in the buying process, the researchers found they were able to predict whether a person would buy a product or pass it up. They concluded, after further analysis of the results, that "the ability of brain activation to predict purchasing would generalize to other purchasing scenarios." *Forbes* heralded the study as a milestone in business, saying it marked the first time researchers have been able "to examine what the brain does while making a purchasing decision." It's not hard to see that we're entering a new era of commerce in which companies will wield far greater influence over our choices than ever before—without our knowing it.

As has been the case so often in the history of information processing, many of the control tools companies are coming to deploy on the Internet have their origins in military research. In the wake of the September 11, 2001, terrorist attacks, the Department of Defense, the Department of Homeland Security, and other US government agencies began investing millions if not billions of dollars into the development of data-mining and analysis technologies that can draw valuable intelligence out of the Net's cloud of data.

Beyond uncovering terrorists, many of these technologies will have applications in the commercial realm. The government has, for instance, funded research at the University of Arizona to develop "stylometry" software that can be used to identify the authors of textual messages appearing on the Internet by analyzing their diction and syntax. Although the research is aimed at "the application of authorship identification techniques to English and Arabic extremist group forum messages," the software could also aid marketers in profiling customers or tracing connections among anonymous product reviews.

None of this means that the computer is about to become purely a control technology. It will always have a dual nature, giving new powers to individuals as well as institutions. We will continue to see advances in information technology that weaken central control, but every disruption will almost certainly be followed by the reassertion of control, whether through legal or technological means. We see this process playing out again today with Napster's heirs, the powerful and highly decentralized peer-to-peer networks that people use to trade movies, software programs, and other large files. Long the realm of libertarians, pirates, and anti-copyright activists, the freewheeling networks have faced a series of lawsuits and police raids in recent years. Now, they're beginning to turn themselves into mainstream media. BitTorrent, the developer of the most popular of the peer-to-peer systems, struck deals in 2006 with several Hollywood studios, agreeing to distribute their movies and TV shows in formats that prevent unauthorized copying. It then turned its home page into a slick online store, required users to sign up for accounts and agree to a long list of "terms of use," and adopted a new name: the BitTorrent Entertainment Network.

History tells us that the most powerful tools for managing the processing and flow of information will be placed in the hands not of ordinary citizens but of businesses and governments. It is their interest—the interest of control—that will ultimately guide the progress and the use of the World Wide Computer.

"AS EVERY MAN goes through life he fills in a number of forms for the record, each containing a number of questions," Alexander Solzhenitsyn wrote in his novel *Cancer Ward*. "A man's answer to one question on one form becomes a little thread, permanently connecting him to the local center of personnel records administration. There are thus hundreds of little threads radiating from every man, millions of threads in all. If these threads were suddenly to become visible, the whole sky would look like a spider's web. . . . Each man, permanently aware of his own invisible threads, naturally develops a respect for the people who manipulate the threads."

As we go about our increasingly digitized lives, the threads that radiate from us are multiplying far beyond anything that even Solzhenitsyn could have imagined in the Soviet Union in the 1960s. Nearly everything we do online is recorded somewhere in the machinery of the World Wide Computer. Every time we read a page of text or click on a link or watch a video, every time we put something in a shopping cart or perform a search, every time we send an email or chat in an instant-messaging window, we are filling in a "form for the record." Unlike Solzhenitsyn's Everyman, however, we're often unaware of the threads we're spinning and how and by whom they're being manipulated. And even if we were conscious of being monitored or controlled, we might not care. After all, we also benefit from the personalization that the Internet makes possible—it makes us more perfect consumers and workers. We accept greater control in return for greater convenience. The spider's web is made to measure, and we're not unhappy inside it.

iGod

I N THE SUMMER of 2004, Google's founders, Larry Page and Sergey Brin, nearly sabotaged their own company. They sat down for a long interview with *Playboy*, and the magazine published the transcript in early August, just days before Google's scheduled debut on the NASDAQ stock exchange. The appearance of the interview roiled Wall Street, as it seemed to violate the Securities and Exchange Commission's prohibition on unauthorized disclosures of information during the "quiet period" leading up to an IPO. Investors feared that the SEC might force the company to cancel its stock offering. But after Google hurriedly distributed a revised prospectus, including the entire text of the *Playboy* interview as an appendix, the SEC cleared the stock sale, and on August 19 Google became a public company.

Lost in the hubbub was the interview itself, which provided a fascinating look into the thoughts and motivations of a pair of brilliant young mathematicians who were about to join the ranks of the world's wealthiest and most powerful businessmen. Toward the end of the interview, Page and Brin gave voice to their deepest ambition. They weren't just interested in perfecting their search engine, they said. What they really looked forward to was melding their technology with the human brain itself. "You want access to as much [information] as possible so you can discern what is most relevant and correct," explained Brin. "The solution isn't to limit the information you receive. Ultimately you want to have the entire world's knowledge connected directly to your mind."

The interviewer was taken aback. "Is that what we have to look forward to?" he asked.

"I hope so," said Brin. "At least a version of that. We probably won't be looking up everything on a computer."

The interviewer probed again: "Is your goal to have the entire world's knowledge connected directly to our minds?"

"To get closer to that—as close as possible," replied Brin. "The smarter we can make the search engine, the better. Where will it lead? Who knows? But it's credible to imagine a leap as great as that from hunting through library stacks to a Google session, when we leap from today's search engines to having the entirety of the world's information as just one of our thoughts."

It wasn't the first time that Brin and Page had talked about their desire to tinker with the human brain—and it wouldn't be the last. In fact, the creation of an artificial intelligence that extends or even replaces the mind is a theme they return to again and again. "Every time I talk about Google's future with Larry Page," reports Steve Jurvetson, a prominent Silicon Valley venture capitalist, "he argues that it will become an artificial intelligence." During a question-and-answer session after a presentation at his alma mater, Stanford University, in May 2002, Page said that Google would fulfill its mission only when its search engine was "AI-complete." "You guys know what that means?" he quizzed the audience of students. "That's artificial intelligence."

In another presentation at Stanford a few months later, Page reiterated the goal: "The ultimate search engine is something as smart as people—or smarter. . . . For us, working on search is a way to work on artificial intelligence." Around the same time, in an interview on public television's *NewsHour,* Brin explained that the "ultimate search engine" would resemble the talking supercomputer HAL in the movie *2001: A Space Odyssey.* "Now, hopefully," said Brin, "it would never have a bug like HAL did where he killed the occupants of the spaceship. But that's what we're striving for, and I think we've made it a part of the way there."

In July 2003, during a talk at a technology conference, Brin and Page went into more detail about their aspiration to use artificial intelligence to make us smarter. Brin suggested, according to a report from a member of the audience, that "wireless brain appliances" might be used to automate the delivery of information. Page elaborated on that idea in a February 2004 interview with Reuters, saying, "On the more exciting front, you can imagine your brain being augmented by Google. For example you think about something and your cell phone could whisper the answer into your ear."

Brin also discussed Google's progress toward its ultimate goal in an interview with *Newsweek* writer Steven Levy. "I think we're pretty far along [with Internet searching] compared to ten years ago," he said. "At the same time, where can you go? Certainly if you had all the world's information directly attached to your brain, or an artificial brain that was smarter than your brain, you'd be better off. Between that and today, there's plenty of space to cover." David Vise relates a similar remark by Brin in his 2005 book *The Google Story.* "Why not improve the brain?" Brin muses at one point. "Perhaps in the future, we can attach a little version of Google that you just plug into your brain."

At a London conference in May 2006, Larry Page again spoke of Google's pursuit of artificial intelligence. "We want to create the ultimate search engine," he said. "The ultimate search engine would understand everything in the world." A year later, in February 2007, he told a group of scientists that Google has a team of employees who are "really trying to build an artificial intelligence and to do it on a large scale." The fulfillment of their goal, he said, is "not as far off as people think."

In taking a transcendental view of information technology, seeing it as a way to overcome what they perceive to be the physical limitations of the human brain, Brin and Page are expressing a desire that has long been a hallmark of the mathematicians and computer scientists who have devoted themselves to the creation of artificial intelligence. It's a desire that, as David Noble notes in *The Religion of Technology*, can be traced all the way back to the seventeenth-century French philosopher René Descartes, who argued that "the body is always a hindrance to the mind in its thinking" and saw in mathematics a model for "pure understanding." The Cartesian ideal runs through the work of mathematicians like George Boole, Alfred North Whitehead, and Alan Turing, whose breakthroughs in algebraic logic set the stage for the modern binary computer.

In her 1979 book *Machines Who Think*, Pamela McCorduck wrote that artificial intelligence promises to provide "an extension of those human capacities we value most." She quoted MIT professor Edward Fredkin's claim that "artificial intelligence is the next step in evolution." Danny Hillis, whose pioneering work in parallel computing paved the way for Google's systems, argued in a 1992 interview that AI could provide a means of remedying man's mental shortcomings, of fixing the "bugs left over history, back from when we were animals," and lead to the creation of beings who are "better than us." In "Reinventing Humanity," a 2006 article, the acclaimed inventor and author Ray Kurzweil predicted that artificial intelligence "will vastly exceed biological intelligence by the mid-2040s," resulting in "a world where there is no distinction between the biological and the mechanical, or between physical and virtual reality."

To most of us, the desire of the AI advocates to merge computers and people, to erase or blur the boundary between man and machine, is troubling. It's not just that we detect in their enthusiasm a disturbing misanthropy—Hillis dismisses the human body as "the monkey that walks around," while Marvin Minsky, the former director of MIT's artificial intelligence program, calls the human brain a "bloody mess of organic matter"—it's also that we naturally sense in their quest a threat to our integrity as freethinking individuals. Even Bill Gates finds the concept discomforting. In a 2005 talk in Singapore, he discussed the possibility of connecting people's bodies and brains directly to computers. One of his Microsoft colleagues, he told the audience, "always says to me, 'I'm ready, plug me in.'" But Gates said that he was wary of the idea: "I don't feel quite the same way. I'm happy to have the computer over there and I'm over here."

In addition to finding the prospect of being turned into computer-enhanced cyborgs unsettling, we also tend to be skeptical of the idea. It seems far-fetched, even ludicrous—like something out of a particularly fanciful piece of science fiction. Here, though, we part company with Gates. In that same speech, he made it clear that he believes the blending of computers and people is inevitable, that we will, in the foreseeable future, come to be augmented by digital processors and software. "We will have those capabilities," he declared. And evidence suggests that Microsoft, like Google, aims to be a pioneer in creating human–computer interfaces for commercial gain. In 2004, the company was granted a patent for a "method and apparatus for transmitting power and data using the human body." In its filing, Microsoft described how it is developing technology that will turn skin into a new kind of electrical conduit, or "bus," that can be used to connect "a network of devices coupled to a single body." It also noted that "the network can be extended by connecting multiple bodies through physical contact [such as] a handshake. When two or more bodies are connected physically, the linked bodies form one large bus over which power and/or communications signals can be transmitted."

Microsoft's patent is just one manifestation of the many corporate and academic research programs that are aimed at merging computers and people and, in particular, at incorporating human beings more fully into the Internet's computing web. A 2006 study sponsored by the British government's Office of Science and Innovation surveyed some of the most promising of these initiatives. In addition to confirming that our bodies are fated to become data-transmission buses—leading to the rise of "computing on the human platform"—the study's authors document the rapid advances taking place in the melding of the real and virtual worlds. New "ambient displays," they write, promise to make computing "ubiquitous," surrounding us with data and software everywhere we go: "In ubiquitous computing, the physical location of data and processing power is not apparent to the user. Rather, information is made available to the user in a transparent and contextually relevant manner." Within ten years, we won't even have to use keystrokes and mouse clicks to tell computers what we want them to do. There will be "new ways of interacting with computers in which delegated systems perform tasks proactively on users' behalf, tuned precisely to the momentary requirements of time and place."

The researchers also predict that the Google founders' dream of a direct link between the brain and the Internet should become a reality by 2020. That's when we're likely to see "the first physical neural interface," providing "a direct connection between a human or animal brain and nervous system and a computer or computer network." At that point, we'll be able "to interact directly with computers by merely thinking." Such a neural interface promises to be a blessing to many people afflicted with severe disabilities. It could help the blind to see and the paralyzed to move. But its applications go well beyond medicine, the researchers note. It also offers the "potential for outside control of human behavior through digital media." We will become programmable, too.

THE INTERNET DOESN'T just connect information-processing machines. It connects people. It connects us with each other, and it connects us with the machines. Our intelligence is as much a part of the power of the World Wide Computer as the intelligence embedded in software code or microchips. When we go online, we become nodes on the Internet. That's not just a metaphor. It's a reflection of the hyperlinked structure that has from the beginning defined the Web and our use of it. The Internet, and all the devices connected to it, is not simply a passive machine that responds to our commands. It's a thinking machine, if as yet a rudimentary one, that actively collects and analyzes our thoughts and desires as we express them through the choices we make while online—what we do, where we go, whom we talk to, what we upload, what we download, which links we click on, which links we ignore. By assembling and storing billions upon billions of tiny bits of intelligence, the Web forms what the writer John Battelle calls "a database of human intentions." As we spend more time and transact more of our commercial and social business online, that database will grow ever wider and deeper. Figuring out new ways for people—and machines—to tap into the storehouse of intelligence is likely to be the central enterprise of the future.

On November 2, 2005, we got a glimpse of what lies ahead for the World Wide Computer when Amazon.com began testing a new service with a strange name: Mechanical Turk. The name, it turned out, was borrowed from an infamous chess-playing "automaton" that was built in 1770 by a Hungarian baron named Wolfgang von Kempelen. The wooden machine, fashioned to look like a Turkish sorcerer sitting in front of a large cabinet, appeared to play chess automatically, using an elaborate system of gears and levers to move the pieces. In its debut, at the Schönbrunn Palace in Vienna, the Mechanical Turk quickly dispatched its first opponent, a Count Cobenzl, to the delight of the assembled courtiers. News of the remarkably intelligent robot spread rapidly, and von Kempelen took the Turk on a tour of Europe, where it defeated a series of famous challengers, including Napoleon Bonaparte and Benjamin Franklin. It was not until years later, after von Kempelen's death, that the hoax was revealed. Hidden inside the cabinet had been a chess master, who used a system of magnets to follow opponents' moves and make his own. The player had been simulating an artificial intelligence.

Amazon's Mechanical Turk accomplishes a similar feat. It "hides" people inside a software program, using them to carry out tasks that computers aren't yet very good at. Say, for example, that a programmer is writing an application that includes, as one of its steps, the identification of buildings in digital photographs—a job that baffles today's computers but is easy for people to do. Using the Mechanical Turk service, the programmer can write a few simple lines of code to tap into the required intelligence. At the designated point in the running of the program, a request to carry out the "human task" automatically gets posted on Amazon's Turk site, where people compete to perform it for a fee set by the programmer.

As Amazon explains on its Web site, Mechanical Turk stands the usual relationship between computers and people on its head:

"When we think of interfaces between human beings and computers, we usually assume that the human being is the one requesting that a task be completed, and the computer is completing the task and providing the results. What if this process were reversed and a computer program could ask a human being to perform a task and return the results?" That's exactly what Mechanical Turk does. It turns people's actions and judgments into functions in a software program. Rather than the machine working for us, we work for the machine.

We play a similar role, without even realizing it, in the operation of Google's search engine. At the heart of that engine is the Page-Rank algorithm that Brin and Page wrote while they were graduate students at Stanford in the 1990s. They saw that every time a person with a Web site links to another site, he is expressing a judgment. He is declaring that he considers the other site important. They further realized that while every link on the Web contains a little bit of human intelligence, all the links combined contain a great deal of intelligence—far more, in fact, than any individual mind could possibly possess. Google's search engine mines that intelligence, link by link, and uses it to determine the importance of all the pages on the Web. The greater the number of links that lead to a site, the greater its value. As John Markoff puts it, Google's software "systematically exploits human knowledge and decisions about what is significant." Every time we write a link, or even click on one, we are feeding our intelligence into Google's system. We are making the machine a little smarter—and Brin, Page, and all of Google's shareholders a little richer.

In Mechanical Turk and the Google search engine, we begin to see the human mind merging into the artificial mind of the World Wide Computer. In both services, people become subservient to the machine. With Mechanical Turk, we're incorporated into a software program, carrying out a small function without being aware of the greater purpose—just as manual laborers became cogs in long assembly lines. In Google's search engine, our contributions are made unconsciously. Brin and Page have programmed their machine to gather the crumbs of intelligence that we leave behind on the Web as we go about our everyday business.

As the computing cloud grows, as it becomes ubiquitous, we will feed ever more intelligence into it. Using global positioning satellites and tiny radio transmitters, it will track our movements through the physical world as meticulously as it today tracks our clicks through the virtual world. And as the types of commercial and social transactions performed through the Internet proliferate, many more kinds of data will be collected, stored, analyzed, and made available to software programs. The World Wide Computer will become immeasurably smarter. The transfer of our intelligence into the machine will happen, in other words, whether or not we allow chips or sockets to be embedded in our skulls.

Computer scientists are now in the process of creating a new language for the Internet that promises to make it a far more sophisticated medium for expressing and exchanging intelligence. In creating Web pages today, programmers have limited options for using codes, or tags, to describe text, images, and other content. The Web's traditional hypertext markup language, or HTML, concentrates on simple formatting commands—on instructing, for instance, a Web browser to put a line of text into italics or to center it on a page. The new language will allow programmers to go much further. They'll be able to use tags to describe the meaning of objects like words and pictures as well as the associations between different objects. A person's name, for instance, could carry with it information about the person's address and job, likes and dislikes, and relationships to other people. A product's name could have tags describing its price, availability, manufacturer, and compatibility with other products.

This new language, software engineers believe, will pave the way for much more intelligent "conversations" between computers on the Internet. It will turn the Web of information into a Web of meaning—a "Semantic Web," as it's usually called. HTML's inventor, Tim Berners-Lee, is also spearheading the development of its replacement. In a speech before the 2006 International World Wide Web Conference in Scotland, he said that "the Web is only going to get more revolutionary" and that "twenty years from now, we'll look back and say this was the embryonic period." He foresees a day when the "mechanisms of trade, bureaucracy and our daily lives will be handled by machines talking to machines."

At the University of Washington's Turing Center, a leading artificial intelligence laboratory, researchers have already succeeded in creating a software program that can, at a very basic level, "read" sentences on Web pages and extract meaning from them—without requiring any tags from programmers. The software, called Text-Runner, scans sentences and identifies the relationships between words or phrases. In reading the sentence "Thoreau wrote *Walden* after leaving his cabin in the woods," for instance, TextRunner would recognize that the verb "wrote" describes a relationship between "Thoreau" and "*Walden.*" As it scans more pages and sees hundreds or thousands of similar constructions, it would be able to hypothesize that Thoreau is a writer and *Walden* is a book. Because TextRunner is able to read at an extraordinary rate—in one test, it extracted a billion textual relationships from 90 million Web pages—it can learn quickly. Its developers see it as a promising prototype of "machine reading," which they define as "the automatic, unsupervised understanding of text" by computers.

Scientists are also teaching machines how to see. Google has been working with researchers at the University of California at San Diego to perfect a system for training computers to interpret photographs and other images. The system combines textual tags describing an image's contents with a statistical analysis of the image. A computer is first trained to recognize an object—a tree, say—by being shown many images containing the object that have been tagged with the description "tree" by people. The computer learns to make an association between the tag and a mathematical analysis of the shapes appearing in the images. It learns, in effect, to spot a tree, regardless of where the tree happens to appear in a given picture. Having been seeded with the human intelligence, the computer can then begin to interpret images on its own, supplying its own tags with ever increasing accuracy. Eventually, it becomes so adept at "seeing" that it can dispense with the trainers altogether. It thinks for itself.

In 1945, the Princeton physicist John von Neumann sketched out the first plan for building an electronic computer that could store in its memory the instructions for its use. His plan became the blueprint for all modern digital computers. The immediate application of von Neumann's revolutionary machine was military—designing nuclear bombs and other weapons—but the scientist knew from the start that he had created a general purpose technology, one that would come to be used in ways that could not be foretold. "I am sure that the projected device, or rather the species of devices of which it is to be the first representative, is so radically new that many of its uses will become clear only after it has been put into operation," he wrote to Lewis Strauss, the future chairman of the Atomic Energy Commission, on October 24, 1945. "Uses which are likely to be the most important are by definition those which we do not recognize at present because they are farthest removed from our present sphere."

We are today at a similar point in the history of the World Wide Computer. We have built it and are beginning to program it, but we are a long way from knowing all the ways it will come to be used. We can anticipate, however, that unlike von Neumann's machine, the World Wide Computer will not just follow our instructions. It will learn from us and, eventually, it will write its own instructions.

GEORGE DYSON, A historian of technology and the son of another renowned Princeton physicist, Freeman Dyson, was invited to Google's headquarters in Mountain View, California, in October 2005 to give a speech at a party celebrating the sixtieth anniversary of von Neumann's invention. "Despite the whimsical furniture and other toys," Dyson would later recall of his visit, "I felt I was entering a 14th-century cathedral — not in the 14th century but in the 12th century, while it was being built. Everyone was busy carving one stone here and another stone there, with some invisible architect getting everything to fit. The mood was playful, yet there was a palpable reverence in the air." After his talk, Dyson found himself chatting with a Google engineer about the company's controversial plan to scan the contents of the world's libraries into its database. "We are not scanning all those books to be read by people," the engineer told him. "We are scanning them to be read by an AI."

The visit inspired Dyson to write an essay for the online journal *Edge,* in which he argues that we've reached a turning point in the history of computing. The computer we use today, von Neumann's computer, uses a physical matrix as its memory. Each bit of data is stored in a precise location on that matrix, with a unique address, and software consists of a set of instructions for finding bits of data at specified addresses and doing something with them. It's a process that, as Dyson explains, "translates informally into 'DO THIS with what you find HERE and go THERE with the result.' Everything depends not only on precise instructions, but on HERE, THERE, and WHEN being exactly defined."

As we know today—and as von Neumann foresaw in 1945—this machine can be programmed to perform an amazing variety of tasks. But it has a fundamental limitation: it can only do what it's told. It depends entirely on the instructions provided by the programmer, and hence it can only perform tasks that a programmer can conceive of and write instructions for. As Dyson writes, "Computers have been getting better and better at providing answers —but only to questions that programmers are able to ask."

That's very different from how living systems, such as our brains, process information. As we navigate our lives, our minds devote most of their time and energy to computing answers to questions that haven't been asked, or at least haven't been asked in precise terms. "In the real world, most of the time," Dyson explains, "finding an answer is easier than defining the question. It's easier to draw something that looks like a cat, for instance, than to describe what, exactly, makes something look like a cat. A child scribbles indiscriminately, and eventually something appears that resembles a cat. A solution finds the problem, not the other way around." What makes us so smart is that our minds are constantly providing answers without knowing the questions. They're making sense rather than performing calculations.

For a machine to demonstrate, or at least simulate, that kind of intelligence, it cannot be restricted to a set of unambiguous instructions for acting on a rigidly defined set of data. It needs to be freed from its fixed memory. It needs to lose its machine-ness and begin acting more like a biological system. That is exactly what's becoming possible as the Internet itself becomes a computer. Suddenly, rather than having a finite set of data arrayed precisely in a matrix, we have a superfluity of data floating around in a great unbounded cloud. We have, to switch metaphors, a primordial soup of information that demands to be made sense of. To do that, we need software that acts more like a sense-making brain than like von Neumann's calculating machine—software with instructions that, as Dyson writes, "say simply 'DO THIS with the next copy of THAT which comes along.'"

We see this new kind of software, in embryonic form, in Google's search engine and in other programs designed to mine information from the Web. Google's engineers recognize that, as Dyson puts it, "a network, whether of neurons, computers, words, or ideas, contains solutions, waiting to be discovered, to problems that need not be explicitly defined." The algorithms of the company's search engine already do a very good job of drawing out of the Internet answers to questions that we pose, even when we phrase our questions in ambiguous terms. We don't always know precisely what we're looking for when we do a Google search, but we often find it nonetheless. If the World Wide Computer is a new kind of computer, then the Google search engine is a preview of the new kind of software that will run on it.

Eric Schmidt has said that the company's ultimate product, the one he's "always wanted to build," would not wait to respond to his query but would "tell me what I should be typing." It would, in other words, provide the answer without hearing the question. The product would be an artificial intelligence. It might even be, to quote Sergey Brin again, "an artificial brain that was smarter than your brain."

AND WHAT OF our brains? As we come to rely ever more heavily on the Internet's vast storehouse of information as an extension of or even a substitute for our own memory, will it change the way we think? Will it alter the way we conceive of ourselves and our relationship to the world? As we put ever more intelligence into the Web, will we, individually, become more intelligent, or less so?

In describing the future of the World Wide Computer—the "Machine," in his terminology—Kevin Kelly writes, "What will most surprise us is how dependent we will be on what the Machine knows—about us and about what we want to know. We already find it easier to Google something a second or third time rather than remember it ourselves. The more we teach this megacomputer, the more it will assume responsibility for our knowing. It will become our memory. Then it will become our identity. In 2015 many people, when divorced from the Machine, won't feel like themselves—as if they'd had a lobotomy."* Kelly welcomes the prospect. He believes that the submergence of our minds and our selves into a greater intelligence will mark the fulfillment of our destiny. The human race, he says, finds itself today at a new beginning, a moment when "the strands of mind, once achingly isolated, have started to come together."

Others are less sanguine about our prospects. In early 2005, the playwright Richard Foreman staged his surrealist drama *The Gods Are Pounding My Head* on a stage at St. Mark's Church in Man-

---

* Kelly's description of man's growing dependency on computers carries a disquieting, if inadvertent, echo of a passage in the notorious manifesto written by Theodore Kaczynski, the Unabomber. "[As] machines become more and more intelligent," Kaczynski wrote, "people will let machines make more of their decisions for them, simply because machine-made decisions will bring better results than man-made ones. Eventually a stage may be reached at which the decisions necessary to keep the system running will be so complex that human beings will be incapable of making them intelligently. At that stage the machines will be in effective control. People won't be able to just turn the machines off, because they will be so dependent on them that turning them off would amount to suicide." What was for Kaczynski a paranoia-making nightmare is for Kelly a vision of utopia.

hattan. It was a bleak work, featuring two exhausted lumberjacks wandering through a wasteland of cultural detritus—a "paper-thin world," as one character puts it—and muttering broken, incoherent sentences. In a note to the audience, Foreman described the inspiration for his "elegiac play." "I come from a tradition of Western culture," he wrote, "in which the ideal (my ideal) was the complex, dense and 'cathedral-like' structure of the highly educated and articulate personality—a man or woman who carried inside themselves a personally constructed and unique version of the entire heritage of the West." He feared, however, that this tradition is fading, that it is being erased as we come to draw more of our sense of the world not from the stores of our memory but from the databases of the Internet: "I see within us all (myself included) the replacement of complex inner density with a new kind of self—evolving under the pressure of information overload and the technology of the 'instantly available.'" As we are emptied of our "inner repertory of dense cultural inheritance," Foreman concluded, we seem to be turning into "pancake people—spread wide and thin as we connect with that vast network of information accessed by the mere touch of a button."

It will be years before there are any definitive studies of the effect of extensive Internet use on our memories and thought processes. But anyone who has spent a lot of time online will likely feel at least a little kinship with Foreman. The common term "surfing the Web" perfectly captures the essential superficiality of our relationship with the information we find in such great quantities on the Internet. The English biologist J. Z. Young, in his Reith Lectures of 1950, collected in the book *Doubt and Certainty in Science,* eloquently described the subtle ways our perceptions, ideas, and language change whenever we begin using a new tool. Our technologies, he explained, make us as surely as we make our technologies. That's been true of the tools we use to process matter and energy, but it's been particularly true of the tools we use to process information, from the map to the clock to the computer.

The medium is not only the message. The medium is the mind. It shapes what we see and how we see it. The printed page, the dominant information medium of the past 500 years, molded our thinking through, to quote Neil Postman, "its emphasis on logic, sequence, history, exposition, objectivity, detachment, and discipline." The emphasis of the Internet, our new universal medium, is altogether different. It stresses immediacy, simultaneity, contingency, subjectivity, disposability, and, above all, speed. The Net provides no incentive to stop and think deeply about anything, to construct in our memory that "dense repository" of knowledge that Foreman cherishes. It's easier, as Kelly says, "to Google something a second or third time rather than remember it ourselves." On the Internet, we seem impelled to glide across the slick surface of data as we make our rushed passage from link to link.

And this is precisely the behavior that the Internet, as a commercial system, is designed to promote. We are the Web's neurons, and the more links we click, pages we view, and transactions we make—the faster we fire—the more intelligence the Web collects, the more economic value it gains, and the more profit it throws off. We feel like "pancake people" on the Web because that's the role we are assigned to play. The World Wide Computer and those who program it have little interest in our exhibiting what Foreman calls "the thick and multi-textured density of deeply evolved personality." They want us to act as hyperefficient data processors, as cogs in an intellectual machine whose workings and ends are beyond us. The most revolutionary consequence of the expansion of the Internet's power, scope, and usefulness may not be that computers will start to think like us but that we will come to think like computers. Our consciousness will thin out, flatten, as our minds are trained, link by link, to "DO THIS with what you find HERE and go THERE with the result." The artificial intelligence we're creating may turn out to be our own.

A HUNDRED YEARS ago, the utility executives and electrical engineers who joined the Jovian Society saw themselves as the architects of a new and more perfect world. To them, God was "the Great Electrician," animating the universe with an invisible but all-powerful spirit. In pursuing their work they were doing His work as well; His designs were their designs. "The idea of electricity," the Jovians announced, is "binding the world together in a body of brotherhood."

Many of the computer scientists and software engineers who are building the great computing grid of the twenty-first century share a similar sense of the importance—and the beneficence—of their work. It's only the metaphor that has changed. God is no longer the Great Electrician. He has become the Great Programmer. The universe is not the emanation of a mysterious spirit. It is the logical output of a computer. "As soon as the universe began, it began computing," writes MIT professor Seth Lloyd in his 2006 book *Programming the Universe.* "Life, language, human beings, society, culture—all owe their existence to the intrinsic ability of matter and energy to process information." "All living creatures are information-processing machines at some level," argues Charles Seife in another, similarly titled 2006 book, *Decoding the Universe.* "In a sense, the universe as a whole is behaving like a giant information processor—a computer."

Our past and our destiny are inscribed in software code. And now, as all the world's computers are wired together into one machine, we have finally been given the opportunity, or at least the temptation, to perfect the code.

Flame and Filament

O
NE OF MAN'S greatest inventions was also one of his most modest: the wick. We don't know who first realized, many thousands of years ago, that fire could be isolated at the tip of a twisted piece of cloth and steadily fed, through capillary action, by a reservoir of wax or oil, but the discovery was, as Wolfgang Schivelbusch writes in *Disenchanted Night*, "as revolutionary in the development of artificial lighting as the wheel in the history of transport." The wick tamed fire, allowing it to be used with a precision and an efficiency far beyond what was possible with a wooden torch or a bundle of twigs. In the process, it helped domesticate us as well. It's hard to imagine civilization progressing to where it is today by torchlight.

The wick also proved an amazingly hardy creation. It remained the dominant lighting technology all the way to the nineteenth century, when it was replaced first by the wickless gas lamp and then, more decisively, by Edison's electricity-fueled incandescent bulb with its glowing metal filament. Cleaner, safer, and even more efficient than the flame it replaced, the lightbulb was welcomed into homes and offices around the world. But along with its many practical benefits, electric light also brought subtle and unexpected changes to the way people lived. The fireplace, the candle, and the oil lamp had always been the focal points of households. Fire was, as Schivelbusch puts it, "the soul of the house." Families would in the evening gather in a central room, drawn by the flickering flame, to chat about the day's events or otherwise pass the time together. Electric light, together with central heat, dissolved that long tradition. Family members began to spend more time in different rooms in the evening, studying or reading or working alone. Each person gained more privacy, and a greater sense of autonomy, but the cohesion of the family weakened.

Cold and steady, electric light lacked the allure of the flame. It was not mesmerizing or soothing but strictly functional. It turned light into an industrial commodity. A German diarist in 1944, forced to use candles instead of lightbulbs during nightly air raids, was struck by the difference. "We have noticed," he wrote, "in the 'weaker' light of the candle, objects have a different, a much more marked profile—it gives them a quality of 'reality.'" This quality, he continued, "is lost in electric light: objects (seemingly) appear much more clearly, but in reality it *flattens* them. Electric light imparts too much brightness and thus things lose body, outline, substance—in short, their essence."

We're still attracted to a flame at the end of a wick. We light candles to set a romantic or a calming mood, to mark a special occasion. We buy ornamental lamps that are crafted to look like candleholders with bulbs shaped as stylized flames. But we can no longer know what it was like when fire was the source of all light. The number of people who remember life before the arrival of Edison's bulb has dwindled to just a few, and when they go they'll take with them all remaining memory of that earlier, pre-electric world. The same will happen, sometime toward the end of this century, with the memory of the world that existed before the computer and the Internet became commonplace. We'll be the ones who bear it away.

All technological change is generational change. The full power and consequence of a new technology are unleashed only when those who have grown up with it become adults and begin to push their outdated parents to the margins. As the older generations die, they take with them their knowledge of what was lost when the new technology arrived, and only the sense of what was gained remains. It's in this way that progress covers its tracks, perpetually refreshing the illusion that where we are is where we were meant to be.

Notes

Prologue: A Doorway in Boston

2   **"IT Doesn't Matter":** For a survey of responses to "IT Doesn't Matter," see Nicholas G. Carr, *Does IT Matter? Information Technology and the Corrosion of Competitive Advantage* (Boston: Harvard Business School Press, 2004), 177–80.

Chapter 1: Burden's Wheel

9   **Henry Burden's story:** See Robert M. Vogel, ed., *A Report of the Mohawk–Hudson Area Survey* (Washington DC: Smithsonian Institution, 1973), 73–95; Louis C. Hunter, *A History of Industrial Power in the United States*, vol. 1, *Waterpower in the Century of the Steam Engine* (Charlottesville: University Press of Virginia, 1979), 569–74; and Paul J. Uselding, "Henry Burden and the Question of Anglo-American Technological Transfer in the Nineteenth Century," *Journal of Economic History* 30 (June 1970), 312–37.

12   **similarities between electricity and computing:** Computing's resemblance to electricity has been noted in the past. In a 1964 essay, Martin Greenberger wrote, "The computing machine is fundamentally an extremely useful device. The service it provides has a kind of universality and generality not unlike that afforded by electric power. Electricity

can be harnessed for any of a wide variety of jobs. . . . Symbolic computation can be applied to an equally broad range of tasks." Expanding on the analogy, Greenberger foresaw the rise of utility computing: "Barring unforeseen obstacles, an on-line interactive computer service, provided commercially by an information utility, may be as commonplace by A.D. 2000 as telephone service is today." In a 1990 article on industrial productivity, Paul A. David noted "the parallel between the modern computer and another general purpose engine, one that figured prominently in what sometimes is called the 'second Industrial Revolution'—namely, the electric dynamo. . . . Computer and dynamo each form the nodal elements of physically distributed (transmission) networks. . . . In both instances, we can recognize the emergence of an extended trajectory of incremental technical improvements, the gradual and protracted process of diffusion into widespread use, and the confluence with other streams of technological innovation, all of which are interdependent features of the dynamic process through which a general purpose engine acquires a broad domain of specific applications." Martin Greenberger, "The Computers of Tomorrow," in Zenon W. Pylyshyn, ed., *Perspectives on the Computer Revolution* (Englewood Cliffs: Prentice Hall, 1970), 390–97; Paul A. David, "The Computer and the Dynamo: A Historical Perspective on the Modern Productivity Paradox," *American Economic Review* 80 (May 1990), 355–61.

15 **general purpose technologies:** The term was introduced in a 1989 paper by Timothy F. Bresnahan and Manuel Trajtenberg, later published as "General Purpose Technologies: 'Engines of Growth'?" *Journal of Econometrics* 65 (1995), 83–108. See also Elhanan Helpman, ed., *General Purpose Technologies and Economic Growth* (Cambridge MA: MIT Press, 1998).

20 **more than 26 million people:** "Global Napster Usage Plummets, But New File-Sharing Alternatives Gaining Ground, Reports Jupiter Media Metrix," ComScore Networks press release, July 20, 2001.

22 **"Western society has accepted":** Lewis Mumford, *The Myth of the Machine*, vol. 2, *The Pentagon of Power* (New York: Harcourt Brace Jovanovich, 1970), 185–86.

Chapter 2: The Inventor and His Clerk

25 **"Why cannot the power"**: Quoted in Neil Baldwin, *Edison: Inventing the Century* (New York: Hyperion, 1995), 104.

26 **"It was not only necessary"**: Quoted in Thomas P. Hughes, *Networks of Power: Electrification in Western Society, 1880–1930* (Baltimore: Johns Hopkins, 1983), 22. Comments Hughes: "Why did Edison so often choose to work on systems? If the inventor created only a component, he remained dependent on others to invent or supply other components. The inventor of components could not have the control over innovation that Edison wanted. [Edison] sought the stimulation for inventive ideas which comes from seeing inadequacies in some components revealed by improvements made in others. Imbalances among interacting components pointed up the need for additional invention" (21–22).

26 **"celebrated as cleanliness"**: Wolfgang Schivelbusch, *Disenchanted Night: The Industrialization of Light in the Nineteenth Century* (Berkeley: University of California Press, 1995), 51.

27 **"barbarous and wasteful"**: Quoted in Baldwin, *Edison*, 137.

27 **"Like the candle and the oil lamp"**: Schivelbusch, *Disenchanted Night*, 56.

28 **"a little globe of sunshine"**: From a *New York Herald* article by Marshall Fox; quoted in Baldwin, *Edison*, 113.

28 **"in a twinkling"**: Quoted in Baldwin, *Edison*, 138.

30 **"a peculiar metabolic make-up"**: Forrest McDonald, *Insull* (Chicago: University of Chicago Press, 1962), 6–8.

31 **"One evening I was going"**: Samuel Insull, *Public Utilities in Modern Life* (Chicago: privately printed, 1924), 185–86.

32 **"From that moment"**: McDonald, *Insull*, 22.

33 **"was so far-fetched"**: Ibid., 54.

34 **"During unrecorded millennia"**: Louis C. Hunter and Lynwood Bryant, *A History of Industrial Power in the United States, 1780–1930*, vol. 3, *The Transmission of Power* (Cambridge MA: MIT Press, 1991), 3.

34   **"Uncounted men and animals":** Ibid., 5.

34   **thousands of such mills:** Louis C. Hunter, *A History of Industrial Power in the United States, 1780–1930*, vol.1, *Waterpower in the Century of the Steam Engine* (Charlottesville: University Press of Virginia, 1979), 2.

36   **"present[ed] a bewildering appearance":** Quoted in Hunter and Bryant, *History of Industrial Power,* vol. 3, 120.

37   **less than 5 percent of the power:** Ibid., 210.

37   **"no one would now think":** Quoted in ibid., 231.

37   **"In the early years":** Ibid., 242.

37   **50,000 power plants, 3,600 central stations:** Amy Friedlander, *Power and Light: Electricity in the US Energy Infrastructure, 1870–1940* (Reston: Corporation for National Research Initiatives, 1996), 51.

39   **"Kemmler Westinghoused":** Richard Moran, "The Strange Origins of the Electric Chair," *Boston Globe,* August 5, 1990.

39   **"The opportunity to get this large power business":** Samuel Insull, *The Memoirs of Samuel Insull* (Polo IL: Transportation Trails, 1992), 79.

40   **Fisk Street dynamos:** Hughes, *Networks of Power,* 211.

42   **a genius at load balancing:** In a speech on November 16, 1916, Insull would explain in remarkable depth the sophistication with which his company optimized its mix of customers, down to the level of individual apartment houses. See Insull, *Public Utilities,* 69–107.

42   **"the historic managerial contributions":** Hughes, *Networks of Power,* 217.

43   **"although isolated plants are still numerous":** The *Electrical World and Engineer* and *Electrical Review and Western Electrician* stories are quoted in ibid., 223.

43   **rising from about 10 kilowatt-hours:** Insull, *Public Utilities,* 70.

44   **utilities' share of total US electricity production:** R. B. DuBoff, *Electric Power in American Manufacturing, 1889–1958* (New York: Arno Press, 1979), 40.

Chapter 3: Digital Millwork

45   **Hollerith's machine:** For a contemporary description of the tabulator, see F. H. Wines, "The Census of 1900," *National Geographic,* January 1900. Wines was the assistant director of the Census Bureau at the time.

47   **"Processing information gathered":** Paul E. Ceruzzi, *A History of Modern Computing* (Cambridge MA: MIT Press, 2003), 16.

48   **dismissed as "foolishness":** Ibid., 14.

49   **"The Utopia of automatic production":** Roddy F. Osborn, "GE and UNIVAC: Harnessing the High Speed Computer," *Harvard Business Review,* July–August 1954.

50   **40,000 reservations and 20,000 ticket sales:** Martin Campbell-Kelly, *From Airline Reservations to Sonic the Hedgehog: A History of the Software Industry* (Cambridge MA: MIT Press, 2003), 41–45.

51   **IT as percentage of capital budget:** US Department of Commerce, *The Emerging Digital Economy,* April 1998, 6 and A1–7.

51   **software spending growth:** Campbell-Kelly, *From Airline Reservations,* 14–15.

51   **global IT expenditures:** John Gantz, "40 Years of IT: Looking Back, Looking Ahead," IDC white paper, 2004.

52   **IBM mainframe rent:** Ceruzzi, *A History of Modern Computing,* 74.

52   **"priesthood of technicians":** Ibid., 77.

56   **capacity utilization of server computers:** Artur Andrzejak, Martin Arlitt, and Jerry Rolia, "Bounding the Resource Savings of Utility Computing Models," Hewlett–Packard Laboratories Working Paper HPL–2002–339, November 27, 2002.

56   **capacity utilization of data storage systems:** Carol Hildebrand, "Why Squirrels Manage Storage Better than You Do," *Darwin,* April 2003; Stephen Foskett, "Real-World Storage Utilization," *Storage,* April 2002.

56   **"To waste a CPU cycle":** Brian Hayes, "The Computer and the Dynamo," *American Scientist* 89 (September–October 2001), 390–94.

56   **"can use up to 100 times as much"**: Gary Shamshoian et al., "High-Tech Means High-Efficiency: The Business Case for Energy Management in High-Tech Industries," Ernest Orlando Lawrence Berkeley National Laboratory white paper, December 20, 2005.

57   **"over the next few years"**: Luiz André Barroso, "The Price of Performance," *ACM Queue* 3 (September 2005), 48–53.

59   **"computing may someday be organized"**: Quoted in David Warsh, *Knowledge and the Wealth of Nations* (New York: Norton, 2006), 351.

60   **circle the globe more than 11,000 times:** Olga Kharif, "The Fiber-Optic 'Glut'—in a New Light," *BusinessWeek Online*, August 31, 2001, http://www.businessweek.com/bwdaily/dnflash/aug2001/nf20010831_396.htm.

60   **"When the network becomes as fast"**: Quoted in George Gilder, "The Information Factories," *Wired*, October 2006.

Chapter 4: Goodbye, Mr. Gates

63   **"The next sea change is upon us"**: Bill Gates, "Internet Software Services," Microsoft memorandum, October 30, 2005.

64   **Google's The Dalles data center:** See Kathy Grey, "Port Deal with Google to Create Jobs," *The Dalles Chronicle*, February 16, 2005.

64   **"massive investment"**: Adam Lashinsky, "Chaos by Design," *Fortune*, October 2, 2006.

64   **"looming like an information-age nuclear plant"**: John Markoff and Saul Hansell, "Google's Not-So-Very-Secret Weapon," *International Herald Tribune*, June 13, 2006.

65   **"a hamburger-and-gas pit stop"**: Laura Oppenheimer, "Faster Than They Can Google It, Newcomers Find The Dalles," *Oregonian*, August 20, 2006.

66   **"the biggest computer in the world"**: Markoff and Hansell, "Google's Not-So-Very-Secret Weapon."

67   **"tens of billions"**: Luiz André Barroso, Jeffrey Dean, and Urs Hölzle,

"Web Search for a Planet: The Google Cluster Architecture," *IEEE Micro* 23 (March–April 2003), 22–28.

67   **One analyst estimates:** Saul Hansell and John Markoff, "A Search Engine That's Becoming an Inventor," *New York Times*, July 3, 2006.

69   **"the biggest mouth":** Erick Schonfeld, "The Biggest Mouth in Silicon Valley," *Business 2.0*, September 2003.

69   **rise of Salesforce.com:** Interview with Marc Benioff by author, November 1, 2006.

72   **McKinsey & Company survey:** Interview with McKinsey's Kishore Kanakamedala and Abhijit Dubey by author, November 22, 2006.

72   **Gartner research:** Robert P. Desisto, Ben Pring, Benoit J. Lheureux, and Frances Karamouzis, "SaaS Delivery Challenges On-Premise Software," Gartner research report, September 26, 2006.

73   **Amazon's utility services:** Interview with Adam Selipsky, Amazon Web Services vice president, by author, October 17, 2006.

74   **"makes it possible for SmugMug":** Amazon Web Services, "Success Story: SmugMug," http://www.amazon.com/b?ie=UTF8&node=206910011&me=A36L942TSJ2AJA.

74   **"It's like having Amazon engineers":** Amazon Web Services, "Success Story: YouOS," http://www.amazon.com/b?ie=UTF8&node=242471011&me=A36L942TSJ2AJA.

74   **"there are times":** Wade Roush, "Servers for Hire," *Technology Review* Web site, September 28, 2006, http://www.technologyreview.com/Biztech/17554/.

78   **an entire virtual data center:** Interview with Bryan Doerr by author, July 29, 2006.

81   **"The environment has changed":** Ray Ozzie, "The Internet Services Disruption," Microsoft memorandum, October 28, 2005.

82   **amounts would be "staggering":** David Kirkpatrick, "Microsoft's Cash Versus Google," CNNMoney.com, May 5, 2006, http://money.cnn.com/2006/05/05/technology/fastforward_fortune/index.htm.

Chapter 5: The White City

85   **Columbian Exposition:** See Reid Badger, *The Great American Fair* (Chicago: Nelson Hall, 1979), and Julie K. Rose, "The World's Columbian Exposition: Idea, Experience, Aftermath," August 1, 1996, http://xroads.virginia.edu/~ma96/WCE/title.html.

86   **"The gleaming lights":** Quoted in Carolyn Marvin, "Dazzling the Multitude: Imagining the Electric Light as a Communications Medium," in Joseph J. Corn, ed., *Imagining Tomorrow: History, Technology, and the American Future* (Cambridge MA: MIT Press, 1986), 204–5.

87   **"One lingered long":** Henry Adams, *The Education of Henry Adams: A Biography* (Boston: Houghton Mifflin, 1918), 342–43.

88   **electric utopianism:** Quotes are from David E. Nye, *Electrifying America: Social Meanings of a New Technology* (Cambridge MA: MIT Press, 1990), 149–50; and Howard P. Segal, "The Technological Utopians," in Corn, ed., *Imagining Tomorrow,* 127–28.

89   **General Electric advertising campaign:** See Nye, *Electrifying America,* 265–70.

90   **"huge industry simply melted away":** Gavin Weightman, *The Frozen-Water Trade* (New York: Hyperion, 2003), 244.

91   **"on the assumption":** David E. Nye, *Consuming Power: A Social History of American Energies* (Cambridge MA: MIT Press, 1998), 141.

92   **"The provision of a whole new system":** Quoted in ibid., 143–44.

94   **"the great transformation of American education":** Claudia Goldin, "Egalitarianism and the Returns to Education During the Great Transformation of American Education," *Journal of Political Economy* 107 (December 1999), S65–S94.

96   **"Here is our poetry":** Ezra Pound, "Patria Mia," *The New Age* 11 (September 19, 1912), 491–92.

96   **"Electricity made possible":** Nye, *Consuming Power,* 157, 164.

98   **1925 Muncie survey:** Ruth Schwartz Cowan, *More Work for Mother* (New York: Basic Books, 1983), 173.

98   **"a domestic engineer":** Edison's comments appeared in the October 1912 issue of *Good Housekeeping*. Quoted in Sheila M. Rothman, *Woman's Proper Place* (New York: Basic Books, 1978), 18.

99   **A photograph of the time:** Reprinted in Nye, *Electrifying America*, 264.

100   **"the labor saved":** Cowan, *More Work for Mother*, 178.

100   **historical studies of housework:** Ibid., 159, 178, 199.

100   **2006 domestic work study:** Valerie A. Ramey and Neville Francis, "A Century of Work and Leisure," National Bureau of Economic Research Working Paper No. 12264, May 2006.

100   **"proletarianization":** Cowan, *More Work for Mother*, 180.

101   **"bring[ing] the home into harmony":** Nye, *Electrifying America*, 252–53.

101   **"The adoption of new technologies":** Ibid., 280–81.

102   **Jovian Society:** See ibid., 161.

Chapter 6: World Wide Computer

105   **epigraph to Part 2:** John M. Culkin, SJ, "A Schoolman's Guide to Marshall McLuhan," *Saturday Review*, March 18, 1967. Culkin is here paraphrasing McLuhan.

107   **"cybernetic meadow":** Brautigan's poem, "All Watched Over by Machines of Loving Grace," was originally published in a collection by the same name in 1967 by the Communication Company in San Francisco. The 1,500 copies were given away. The poem is reprinted in *Richard Brautigan's Trout Fishing in America, The Pill Versus the Springhill Mine Disaster, and In Watermelon Sugar* (New York: Mariner Books, 1989).

107   **"be used for exchanging messages":** Defense Advanced Research Projects Agency, Information Processing Techniques Office, "ARPANET Completion Report," January 4, 1978, II–7, II–8, and III–14.

108 **As Fred Turner describes:** Fred Turner, *From Counterculture to Cybercul-ture: Stewart Brand, the Whole Earth Network, and the Rise of Digital Uto-pianism* (Chicago: University of Chicago, 2006). In explaining the affinity between hippies and computers, Turner writes: "To a generation that had grown up in a world beset by massive armies and by the threat of nuclear holocaust, the cybernetic notion of the globe as a single, interlinked pattern of information was deeply comforting; in the invisible play of information, many thought they could see the possibility of global harmony" (5).

109 **"Ready or not":** Stewart Brand, "Spacewar: Fanatic Life and Symbolic Death Among the Computer Bums," *Rolling Stone,* December 7, 1972.

109 **"web of knowledge":** Tim Berners-Lee, *Weaving the Web: The Original Design and Ultimate Destiny of the World Wide Web by Its Inventor* (New York: Harper-Collins, 1999), 2, 162.

109 **"Heavenly City":** Michael Benedikt, "Introduction," in Benedikt, ed., *Cyberspace: First Steps* (Cambridge MA: MIT Press, 1991), 14–16.

109 **"opens up a space":** Nicole Stenger, "Mind Is a Leaking Rainbow," in Benedikt, ed., *Cyberspace,* 49–58.

109 **"the most transforming technological event":** John Perry Barlow et al., "What Are We Doing On-Line?" *Harper's,* August 1995.

109 **"the new home of Mind":** John Perry Barlow, "A Declaration of the Indepen-dence of Cyberspace," February 8, 1996, http://homes.eff.org/~barlow/ Declaration-Final.html.

110 **dominance of commercial sites:** Matthew Gray, "Internet Statistics: Growth and Usage of the Web and the Internet," 1996, http://www.mit. edu/people/mkgray/net/web-growth-summary.html.

112 **service monopolies:** While the modularity of information technology encourages a diversity of suppliers, other forces in the nascent utility-computing industry, such as the heavy capital investments required to build data centers and the economies of scale and scope in delivering online services and advertising, encourage the formation of monopolies. In a possible preview of what's to come, the US Congress launched in July 2007 an investigation of Google's power to squelch competition.

117 **"goes well beyond simple communication":** CERN, "What Is the

Grid?" http://gridcafe.web.cern.ch/gridcafe/whatisgrid/whatis.html. Similar grids of home PCs and even PlayStation gaming consoles are being used for other philanthropic purposes. Stanford University's Folding@Home program, for example, draws on the spare capacity of more than 50,000 PCs and consoles to model protein molecules that can cause diseases. Inspired by the success of the program, Sony is considering setting up a PlayStation grid to carry out commercial computing jobs for pharmaceutical companies and other paying clients.

122   **"information is a commodity"**: James R. Beniger, *The Control Revolution: Technological and Economic Origins of the Information Society* (Cambridge MA: Harvard University Press, 1986), 408.

123   **"hi-tech computers on wheels"**: Bobbie Johnson, "Microsoft Unveils Sync In-Car Computer System," *Guardian Unlimited*, January 8, 2007, http://technology.guardian.co.uk/news/story/0,,1985195,00.html.

123   **"has resulted in several new classes"**: Center for Embedded Networked Sensing, *Annual Progress Report*, May 1, 2006, 5.

124   **"gargantuan Machine"**: Kevin Kelly, "We Are the Web," *Wired*, August 2005.

124   **"say they 'feel as strongly'"**: Center for the Digital Future, "Online World as Important to Internet Users as Real World?" USC–Annenberg School for Communication press release, November 29, 2006.

125   **"The simple faith in progress"**: Norbert Wiener, *The Human Use of Human Beings* (New York: Doubleday Anchor, 1954), 47.

Chapter 7: From the Many to the Few

128   **"Hi, YouTube"**: The Hurley–Chen video can be seen at http://www.youtube.com/watch?v=QCVxQ_3Ejkg.

131   **"may be the fastest-growing product"**: Mary Meeker and David Joseph, "The State of the Internet, Part 3: The World's Information Is Getting Organized & Monetized," presentation at Web 2.0 Summit, San Francisco, November 8, 2006.

131   **"one Skype employee"**: Allan Martinson, *Äripäev*, September 14,

2005, translated and reprinted on Skype website, September 19, 2005, http://share.skype.com/sites/en/2005/09/ebay_a_view_from _estonian_medi.html.

132   **"It amazes me":** Markus Frind, "Small Companies & Google Adsense Is the Future," *Paradigm Shift* blog, June 7, 2006, http://plentyoffish.wordpress .com/2006/06/07/small-companies-google-adsense-is-the-future/.

134   **newspaper job losses:** "News Staffs Shrinking While Minority Presence Grows," American Society of Newspaper Editors press release, April 12, 2005.

135   **"Web 2.0 and the Net":** Philip Dawdy, "Love American Style: Web 2.0 and Narcissism," *Furious Seasons* blog, December 27, 2006, http://www.furious seasons.com/archives/2006/12/love_american_style_web_20_and _narcissism_1.html.

135   **"The Internet is the wave":** Floyd Norris, "Looking for a Paycheck? Don't Look to the Internet," *New York Times*, February 10, 2007.

136   **"Substitution of machinery":** David H. Autor, Frank Levy, and Richard J. Murnane, "The Skill Content of Recent Technological Change: An Empirical Exploration," *Quarterly Journal of Economics* 118 (November 2003), 1279–333.

139   **"cyber-sweatshop":** Lisa Margonelli, "Inside AOL's Cyber-Sweatshop," *Wired*, October 1999.

140   **"First, the physical machinery":** Yochai Benkler, *The Wealth of Networks: How Social Production Transforms Markets and Freedom* (New Haven: Yale University Press, 2006), 105–6.

140   **"networked information economy":** Ibid., 468–73.

141   **"Unrestricted by physical distance":** Richard Barbrook, "The Hi-Tech Gift Economy," *First Monday* 3, no. 12 (1998), http://firstmonday.org/ issues/issue3_12/barbrook/index.html. See also Barbrook, "Giving Is Receiving," *Digital Creativity* 14 (June 2003), 91–94.

142   **"With less than 10 people":** Steven Levy and Brad Stone, "The New Wisdom of the Web," *Newsweek*, April 3, 2006.

142   **"glut of images"**: Sion Touhig, "How the Anti-Copyright Lobby Makes Big
Business Richer," *Register*, December 29, 2006, http://www.theregister
.co.uk/2006/12/29/photojournalism_and_copyright/.

144   **skewing of American incomes:** Thomas Piketty and Emmanuel Saez,
"Income Inequality in the United States, 1913–2002," working paper,
November 2004, and accompanying Excel spreadsheet, http://elsa
.berkeley.edu/~saez/TabFig2004prel.xls. See also Ian Dew-Becker and
Robert J. Gordon, "Where Did the Productivity Growth Go? Inflation
Dynamics and the Distribution of Income," National Bureau of Eco-
nomic Research Working Paper No. 11842, December 2005.

144   **comparison of top-executive and average-worker compensation:**
Carola Frydman and Raven E. Saks, "Historical Trends in Executive
Compensation, 1936–2003," MIT Sloan School of Management working
paper, November 15, 2005.

145   **"plutonomy":** Ajay Kapur, Niall Macleod, and Narendra Singh, "Plutonomy: Buying Luxury, Explaining Global Imbalances," Citigroup
industry note, October 16, 2005.

145   **"There are assembly lines":** Jagdish Bhagwati, "Technology, Not Globalization, Is Driving Down Wages," *Financial Times*, January 4, 2007.

145   **"the influence of globalization on inequality":** Ben S. Bernanke, "The
Level and Distribution of Economic Well-Being," speech before the
Greater Omaha Chamber of Commerce, Omaha, Nebraska, February 6,
2007.

146   **"millions of ordinary people":** Chris Anderson, *The Long Tail: Why the
Future of Business Is Selling Less of More* (New York: Hyperion, 2006), 65.

Chapter 8: The Great Unbundling

150   **"Once the most popular fare":** Chris Anderson, "Nobody Goes There
Anymore—It's Too Crowded," *The Long Tail* blog, December 10,
2006, http://www.longtail.com/the_long_tail/2006/12/nobody_goes
_the.html.

151 **"the tyranny of lowest-common-denominator fare":** Anderson, *The Long Tail*, 16, 159.

151 **decline in newspaper readership:** Julia Angwin and Joseph T. Hallinan, "Newspaper Circulation Continues Decline, Forcing Tough Decisions," *Wall Street Journal*, May 2, 2005; Newspaper Association of America, "Newspapers By the Numbers," http://www.naa.org/thesource/14.asp #number; "Big Metros Show Severe Decline in Latest Circ Result," *Editor & Publisher*, October 30, 2006; Newspaper Association of America, "Readership Statistics," http://www.naa.org/readershippages/research-and-readership/readership-statistics.aspx.

152 **shift to online news:** John B. Horrigan, "Online News: For Many Broadband Users, the Internet Is a Primary News Source," Pew Internet & American Life Project report, March 22, 2006.

152 **decline in advertising revenues:** David Hallerman, "U.S. Online Ad Spending: Peak or Plateau?" eMarketer analyst report, October 2006.

152 **shift in classified ad sales:** Catherine Holahan, "*Journal* Blazes Newspapers' New Trail," *BusinessWeek*, January 3, 2007.

153 **"For virtually every newspaper":** Ibid.

153 **growth in online readership:** "Newspaper Online Readership Jumped in Q4, Setting Records," *Editor & Publisher*, February 7, 2007.

155 **"I'd be moving to the Web":** Jeff Smith, "Newspapers Needn't Fear Craigslist," *Rocky Mountain News*, March 6, 2006.

156 **"How do we create high-quality content":** Martin Nisenholtz, comments at the Online Publishers Association Conference, London, March 2, 2006.

156 ***The Times* of London training:** Aaron O. Patrick, "Google This: U.K. Papers Vie to Buy Search Terms," *Wall Street Journal*, January 12, 2007.

156 **"you can't really avoid the fact":** David Carr, "24-Hour Newspaper People," *New York Times*, January 15, 2007.

157 **"bundling of the world's computers":** Daniel Akst, "Unbundles of Joy," *New York Times*, December 11, 2005.

157   **"to pay for detritus"**: William M. Bulkeley, "The Internet Allows Consumers to Trim Wasteful Purchases," *Wall Street Journal*, November 29, 2006.

157   **Schelling's experiment**: Thomas C. Schelling, "Dynamic Models of Segregation," *Journal of Mathematical Sociology* 1 (1971), 143–86.

159   **"Social realities are fashioned"**: Mark Buchanan, *Nexus: Small Worlds and the Groundbreaking Science of Networks* (New York: Norton, 2002), 186.

159   **"While the politicians struggle"**: Nicholas Negroponte, *Being Digital* (New York: Vintage, 1996), 230.

160   **"orders your search results"**: Google Web site, http://www.google.com/support/bin/answer.py?answer=26651&topic=9005.

161   **"audio-fingerprinting"**: Michael Fink, Michele Covell, and Shumeet Baluja, "Social- and Interactive-Television Applications Based on Real-Time Ambient-Audio Identification," paper presented at Euro ITV Conference, Athens, Greece, May 2006.

161   **"100% of a user's data"**: Eric Schmidt, presentation and presentation notes for Google Analyst Day 2006, March 2, 2006.

162   **"Global Village or Cyber-Balkans?"**: Marshall Van Alstyne and Erik Brynjolfsson, "Global Village or Cyber-Balkans? Modeling and Measuring the Integration of Electronic Communities," *Management Science* 51 (June 2005), 851–68.

163   **"Divided They Blog"**: Lada Adamic and Natalie Glance, "The Political Blogosphere and the 2004 U.S. Election: Divided They Blog," paper presented at the Third International Workshop on Link Discovery, Chicago, August 21, 2005.

164   **"Only a handful of sites"**: Matthew Hindman, presentation at Annenberg School for Communication Hyperlinked Society Conference, Philadelphia, June 9, 2006.

164   **"ideological amplification" experiment**: David Schkade, Cass R. Sunstein, and Reid Hastie, "What Really Happened on Deliberation Day?"

AEI–Brookings Joint Center for Regulatory Studies Working Paper 06–19, July 2006.

165 **"When like-minded people cluster":** Cass R. Sunstein, *Infotopia: How Many Minds Produce Knowledge* (New York: Oxford University Press, 2006), 58, 92–93.

165 **"Individuals empowered":** Van Alstyne and Brynjolfsson, "Global Village or Cyber-Balkans?"

Chapter 9: Fighting the Net

170 **"websites like Google Earth":** Jasper Copping, "Insurgents 'Using Google Earth,'" *Daily Telegraph*, December 17, 2005.

170 **"now have the maps":** Thomas Harding, "Terrorists 'use Google maps to hit UK troops,'" *Daily Telegraph*, January 13, 2007.

170 **"good and bad":** Ibid.

170 **electric-shock transmitters:** See Darius Rejali, "Electricity: The Global History of a Torture Technology," Reed College Web site, http://academic.reed.edu/poli_sci/faculty/rejali/rejali/articles/History_of _Electric_Torture.htm.

171 **Dark Web Portal statistics:** Frank Bures, "Global Jihad Online," *Wired*, December 2006.

172 **Information Operations Roadmap:** US Department of Defense report, October 30, 2003.

173 **"spam will be solved":** Todd Bishop, "Is Gates' Prediction on Spam a Bust?" *Seattle Post–Intelligencer,* January 23, 2006.

174 **increase in spam:** Gregg Keizer, "Spam Sets Record, Accounts for 94% of E-mail," *Information Week*, January 10, 2007; "Spam Volume Keeps Rising," CNET, September 1, 2004, http://news.com.com/ Spam+volume+keeps+rising/2100-1032_3-5339257.html; Gregg Keizer, "Spam Volume Jumps 35% in November," *Information Week*, December 21, 2006.

175 **"We are losing this war"**: John Markoff, "Attack of the Zombie Computers Is a Growing Threat, Experts Say," *New York Times*, January 7, 2007.

176 **botnet attack on Blue Security:** See Scott Berinato, "Attack of the Bots," *Wired*, November 2006.

176 **botnet attack on Estonia:** "A Cyber-riot," *Economist*, May 10, 2007; Jose Nazario, "Estonian DDoS Attacks—A Summary So Far," Arbor Networks Web site, May 17, 2007, http://asert.arbornetworks.com/2007/05 /estonian-ddos-attacks-a-summary-to-date.

177 **"pandemic"; "It's as bad"**: Tim Weber, "Criminals May Overwhelm the Web," BBC News, January 25, 2007, http://news.bbc.co.uk/2/hi/ business/6298641.stm.

177 **"some 20 nations"**: Berinato, "Attack of the Bots."

177 **"The IT infrastructure"**: President's Information Technology Advisory Committee, "Cyber Security: A Crisis of Prioritization," February 2005.

177 **"Today, virtually every sector"**: Testimony of Dr. F. Thomson Leighton before the Committee on Science of the US House of Representatives, Thursday, May 12, 2005.

178 **Silicon Valley electricity meeting:** Sarah Jane Tribble, "Tech Firms Worried About Energy Shortages," *Mercury News*, December 6, 2006; David Needle, "Is There a Moral Obligation to Be Energy Efficient?" *Server Watch*, December 8, 2006.

179 **"I haven't experienced"**: Associated Press, "Quake Outage May Be Sign of Fragile Network," *Los Angeles Times*, December 29, 2006.

179 **"the Net is still working"**: Weber, "Criminals May Overwhelm the Web."

180 **"We might just be at the point"**: David Talbot, "The Internet Is Broken," *Technology Review*, December 2005–January 2006.

180 **"small companies are beginning"**: Stephen Craike and John Merryman, "Offshoring Datacenters? You Might Want to Think About That," *CIO Update*, October 5, 2006.

181   **France's BlackBerry ban:** Molly Moore, "Memo to French Officials: Beware the BlackBerry," *Washington Post*, June 21, 2007.

182   **the Net's "real estate":** For a thorough discussion of Internet governance and its relationship to the network's technical structure, see Milton L. Mueller, *Ruling the Root* (Cambridge MA: MIT Press, 2002). Noting that the Net has since the early 1990s served as "a reference point for public discourse about utopia," Mueller says that today "the world is starting to close in on cyberspace." He argues that there's "a life cycle in the evolution of technical systems. Systems that create new resources and new arenas of economic and social activity can escape institutional regimes and create moments of disequilibrating freedom and social innovation. But eventually a new equilibrium is established. The absolute freedom of a global common pool becomes too costly to maintain" (265–67).

183   **"the establishment of an arbitration":** European Union, "Proposal for Addition to Chair's Paper Sub-Com A Internet Governance on Paragraph 5, 'Follow-up and Possible Arrangements,'" Document WSIS–II/PC–3/DT/21-E, September 30, 2005.

183   **"any Internet governance approach":** World Summit on the Information Society, "Tunis Agenda for the Information Society," Document WSIS–05/TUNIS/DOC/6(Rev. 1)–E, November 18, 2005.

183   **"ranging from the United States's":** Jack Goldsmith and Timothy Wu, *Who Controls the Internet? Illusions of a Borderless World* (New York: Oxford University Press, 2006), 184.

184   **"a very high cost":** Carlota Perez, "Technological Revolutions, Paradigm Shifts, and Socio-Institutional Change," in Erik Reinert, ed., *Globalization, Economic Development and Inequality: An Alternative Perspective* (Cheltenham: Edward Elgar, 2004), 241. See also Perez's *Technological Revolutions and Financial Capital* (Cheltenham: Edward Elgar, 2002).

Chapter 10: A Spider's Web

185   **Thelma Arnold article:** Michael Barbaro and Tom Zeller Jr., "A Face Is Exposed for AOL Searcher No. 4417749," *New York Times*, August 9, 2006.

185 **"Having the AOL data":** Katie Hafner, "Researchers Yearn to Use AOL Logs, But They Hesitate," *New York Times*, August 23, 2006.

186 **"a couple of hours":** David Gallagher, email to author, January 16, 2007.

186 **AOL subscriber searches:** Declan McCullagh, "AOL's Disturbing Glimpse into Users' Lives," CNET, August 7, 2006, http://news.com .com/2100-1030_36103098.html.

187 **Amazon.com wish list experiment:** Tom Owad, "Data Mining 101: Finding Subversives with Amazon Wishlists," *Apple Fritter* Web site, January 4, 2006, http://www.applefritter.com/bannedbooks; Owad, emails to author, January 2007.

189 **University of Minnesota privacy experiment:** Dan Frankowski et al., "You Are What You Say: Privacy Risks of Public Mentions," paper presented at SIGIR '06, Seattle, August 6–11, 2006.

190 **"You have zero privacy":** Polly Sprenger, "Sun on Privacy: 'Get Over It,'" Wired News, January 26, 1999, http://www.wired.com/politics/law/ news/1999/01/17538.

191 **"collection of visionaries":** Editors, "Wrapping Up: Internet Liberation," *Cato Unbound*, January 26, 2006, http://www.cato-unbound.org/2006/ 01/26/the-editors/wrapping-up-internet-liberation/.

191 **"The Internet's output":** Clay Shirky, "Andrew Keen: Rescuing 'Luddite' from the Luddites," *Many-to-Many*, July 9, 2007, http://many .corante.com/archives/2007/07/09/andrew_keen_rescuing_luddite _from_the_luddites.php.

191 **"The Web is a world":** David Weinberger, *Small Pieces Loosely Joined* (New York: Perseus, 2002), 25.

193 **"The producer can no longer embrace":** Quoted in Beniger, *Control Revolution*, 11.

194 **"bureaucratic administration did not begin":** Ibid., 14.

195 **"Microprocessor and computer technologies":** Ibid., vii.

196 **Licklider's paper:** J. C. R. Licklider, "Man–Computer Symbiosis," *IRE Transactions on Human Factors in Electronics*, March 1960, 4–11.

196   **"information processing and flows"**: Beniger, *Control Revolution*, 433–34.

197   **"a Biblical plague"**: Ceruzzi, *A History of Modern Computing*, 293.

197   **"Local networking took"**: Ibid., 295.

198   **"are the enemy of bureaucracy"**: Alexander R. Galloway, *Protocol: How Control Exists After Decentralization* (Cambridge MA: MIT Press, 2004), 29.

198   **"Governments of the Industrial World"**: Barlow, "A Declaration of the Independence of Cyberspace."

198   **"a new apparatus of control"**: Galloway, *Protocol*, 3.

199   **Yahoo's political education**: See Goldsmith and Wu, *Who Controls the Internet?* 1–10.

201   **"Strengthening network culture construction"**: Mure Dickie, "China's President Urges Party Officials to 'Purify' Internet," *Financial Times*, January 26, 2007.

201   **199 data-mining programs**: Ellen Nakashima and Alec Klein, "Daylight Sought for Data Mining," *Washington Post*, January 11, 2007; Michael J. Sniffen, "Expert Says Data-Mining Outstrips Federal Control," *Modesto Bee*, January 12, 2007.

201   **NSA data-mining program**: See Eric Lichtblau and James Risen, "Spy Agency Mined Vast Data Trove, Officials Report," *New York Times*, December 24, 2005, and Ryan Singel, "Whistle-Blower Outs NSA Spy Room," *Wired News*, April 7, 2006, http://www.wired.com/science/discoveries/news/2006/04/70619.

202   **BlackBerry addiction**: Katherine Rosman, "BlackBerry Orphans," *Wall Street Journal*, December 8, 2006.

203   **IBM employee modeling**: Stephen Baker, "Math Will Rock Your World," *Business Week*, January 23, 2006.

203   **Google employee questionnaire**: Saul Hansell, "Google Answer to Filling Jobs Is an Algorithm," *New York Times*, January 3, 2007.

204   **"advertising-funded search engines"**: Sergey Brin and Larry Page,

"The Anatomy of a Large-Scale Hypertextual Web Search Engine," Stanford University working paper, 1998.

205 **"MySpace can be viewed"**: Wade Roush, "Fakesters," *Technology Review*, November–December 2006.

206 **"make lots of spearheads"**: "The Ultimate Marketing Machine," *Economist*, July 6, 2006.

207 **"the diversity of [a] user's"**: Jian Hu et al., "Demographic Prediction Based on User's Browsing Behavior," paper presented at the International World Wide Web Conference, Banff, Alberta, May 12, 2007.

207 **"the ability of brain activation"**: Brian Knutson et al., "Neural Predictors of Purchases," *Neuron* 53 (January 2007), 147–57.

207 **"to examine what the brain does"**: Elisabeth Eaves, "This Is Your Brain on Shopping," *Forbes,* January 5, 2007.

208 **"stylometry" study**: Ahmed Abbasi and Hsinchun Chen, "Applying Authorship Analysis to Extremist-Group Web Forum Messages," *IEEE Intelligent Systems* 20 (September–October 2005), 67–75.

209 **"As every man goes through life"**: Alexander Solzhenitsyn, *Cancer Ward* (New York: Farrar, Straus and Giroux, 1969), 192.

Chapter 11: iGod

201 *Playboy* **interview**: David Sheff, "Google Guys," *Playboy,* September 2004.

212 **"Every time I talk"**: Steve Jurvetson, "Thanks for the Memory" (comments), *The J Curve* blog, January 9, 2005, http://jurvetson.blogspot.com/2005/01/thanks-for-memory.html#c110568424074198024.

212 **"AI-complete"**: Larry Page, presentation at Stanford University, May 1, 2002, http://edcorner.stanford.edu/authorMaterialInfo.html?mid=1091&author=149. The term "AI-complete" is used by computer scientists to refer to an (as yet hypothetical) application of artificial intelligence that is indistinguishable from human intelligence.

212 **"The ultimate search engine"**: Rachael Hanley, "From Googol to Google," *Stanford Daily,* February 12, 2003.

213   **"it would never have a bug like HAL did":** Spencer Michaels, "The Search Engine that Could," *NewsHour with Jim Lehrer,* November 29, 2002.

213   **"wireless brain appliances":** Chris Gulker, July 15, 2003, http://www .gulker.com/2003/07/15.html.

213   **"On the more exciting front":** Reuters, "Google Co-founder Bugged by IPO Speculation," CNET, February 27, 2004, http://news.com .com/Google+co-founder+bugged+by+IPO+speculation/2100-1032 _3-5166351.html.

213   **"I think we're pretty far along":** Steven Levy, "All Eyes on Google," *Newsweek*, March 29, 2004.

213   **"Why not improve the brain?"** David Vise, *The Google Story* (New York: Delacorte, 2005), 292.

213   **"We want to create":** Richard Wray, "Google Users Promised Artificial Intelligence," *Guardian*, May 23, 2006.

214   **"really trying to build artificial intelligence":** Larry Page, remarks before the annual meeting of the American Association for the Advancement of Science, San Francisco, February 16, 2007. A video of the talk is available at http://www.aaas.org/meetings/Annual_Meeting/2007 _San_Fran/02_PE/pe_01_lectures.shtml.

214   **The Cartesian ideal:** David F. Noble, *The Religion of Technology: The Divinity of Man and the Spirit of Invention* (New York: Knopf, 1997), 143–71.

214   **"an extension of those human capacities":** Pamela McCorduck, *Machines Who Think*, 2nd ed. (Wellesley: A. K. Peters, 2004), 135.

214   **"artificial intelligence is the next step":** Ibid., 401.

214   **"bugs left over history":** Steven Levy, "A–Life Nightmare," *Whole Earth Review*, Fall 1992.

214   **"will vastly exceed biological intelligence":** Ray Kurzweil, "Reinventing Humanity," *Futurist*, March–April 2006. See also Kurzweil's *The Singularity Is Near: When Humans Transcend Biology* (New York: Viking, 2005).

215 **"the monkey that walks around":** Levy, "A–Life Nightmare."

215 **"bloody mess":** Quoted by Christopher Lasch, "Chip of Fools," *New Republic,* August 13 and 20, 1984.

215 **"always says to me":** Rohan Sullivan, "Gates Says Technology Will One Day Allow Computer Implants—But Hardwiring's Not for Him," *Technology Review,* August 16, 2005, http://www.technologyreview .com/Wire/14556/.

215 **Microsoft patent:** United States Patent No. 6,754,472.

216 **British government innovation survey:** Institute for the Future, "Delta Scan: The Future of Science and Technology, 2005–2055," Stanford University Web site, 2006, http://humanitieslab.stanford.edu/2/Home.

217 **"database of human intentions":** John Battelle, *The Search: How Google and Its Rivals Rewrote the Rules of Business and Transformed Our Culture* (New York: Portfolio, 2005), 1–17.

218 **Mechanical Turk:** For an account of the original chess-playing machine, see Tom Standage, *The Turk: The Life and Times of the Famous Eighteenth-Century Chess-Playing Machine* (New York: Walker Books, 2002). A description of the Amazon service can be found at http://www .mturk.com/mturk/welcome. See also Nicholas Carr, "Amazon Patents Cybernetic Mind-Meld," *Rough Type* blog, April 4, 2007, http://www .roughtype.com/archives/2007/04/amazon_patents.php.

219 **Google PageRank:** See Brin and Page, "Anatomy of a Search Engine."

219 **"systematically exploits human knowledge":** John Markoff, "Entrepreneurs See a Web Guided by Common Sense," *New York Times,* November 12, 2006.

221 **"the Web is only going to get":** Victoria Shannon, "A 'More Revolutionary' Web," *International Herald Tribune,* May 24, 2006.

221 **"mechanisms of trade":** Berners-Lee, *Weaving the Web,* 158.

221 **TextRunner:** See Oren Etzioni, Michele Banko, and Michael J. Cafarella, "Machine Reading," American Association for Artificial Intelligence working paper, 2006.

221   **teaching computers to see:** Gustavo Carneiro et al., "Supervised Learning of Semantic Classes for Image Annotation and Retrieval," *IEEE Transactions on Pattern Analysis and Machine Intelligence* 29 (March 2007), 394–410.

222   **von Neumann's machine:** John von Neumann described his plan for what he called "a very high-speed automatic digital computing system" in a June 30, 1945, report for the US Army Ordnance Department titled "First Draft of a Report on the EDVAC." (The design was inspired by discussions he'd had with J. Presper Eckert and John Mauchly.) In the report, von Neumann drew an analogy between the vacuum tubes that would serve as the core processing elements of his machine and "the neurons of the higher animals."

222   **"Uses which are likely to be":** Quoted in George Dyson, "Turing's Cathedral," *Edge*, October 24, 2005, http://www.edge.org/3rd_culture/ dyson05/dyson05_index.html.

223   **"Despite the whimsical furniture":** Dyson, "Turing's Cathedral." Dyson's article is best read as a coda to his book *Darwin Among the Machines* (Cambridge: Perseus, 1997).

224   **finding answers to unasked questions:** In his 1960 paper "Man–Computer Symbiosis," J. C. R. Licklider made a similar observation about the limits of traditional computers: "Present-day computers are designed primarily to solve preformulated problems or to process data according to predetermined procedures. The course of the computation may be conditional upon results obtained during the computation, but all the alternatives must be foreseen in advance. (If an unforeseen alternative arises, the whole process comes to a halt and awaits the necessary extension of the program.) The requirement for preformulation or predetermination is sometimes no great disadvantage. It is often said that programming for a computing machine forces one to think clearly, that it disciplines the thought process. If the user can think his problem through in advance, symbiotic association with a computing machine is not necessary. However, many problems that can be thought through in advance are very difficult to think through in advance. They would be easier to solve, and they could be solved faster, through an intuitively guided

trial-and-error procedure in which the computer cooperated, turning up flaws in the reasoning or revealing unexpected turns in the solution. Other problems simply cannot be formulated without computing-machine aid. [French mathematician Henri] Poincaré anticipated the frustration of an important group of would-be computer users when he said, 'The question is not, "What is the answer?" The question is, "What is the question?"' One of the main aims of man–computer symbiosis is to bring the computing machine effectively into the formulative parts of technical problems."

225 **"always wanted to build":** Eric Schmidt, presentation at Google Press Day, May 10, 2006.

226 **"What will most surprise us":** Kelly, "We Are the Web."

227 **"I come from a tradition":** Richard Foreman, "The Pancake People, or, 'The Gods Are Pounding My Head,'" *Edge*, March 8, 2005, http://www .edge.org/3rd_culture/foreman05/foreman05_index.html.

227 **Our technologies make us:** Young describes, for instance, the profound effect that the invention of the clock had on our sense of the world: "In our civilization the emphasis on a steady flow of time developed mainly with the invention of good clocks. This conception provided an important part of the framework of the new view of a material world that man created in the seventeenth century. Moreover, natural events were compared with clocks. The heavens and the human body were said to function like clockwork. In fact, use of this model was largely responsible for the practice of speaking of the 'working' of any system—a practice still deeply ingrained in us." J. Z. Young, *Doubt and Certainty in Science: A Biologist's Reflections on the Brain* (London: Oxford University Press, 1951), 103.

228 **"its emphasis on logic":** Neil Postman, *Technopoly: The Surrender of Culture to Technology* (New York: Vintage, 1993), 16.

229 **"The idea of electricity":** Nye, *Electrifying America*, 161.

229 **God as technologist:** Before God was the Great Programmer or the Great Electrician, He was the Great Clockmaker, a metaphor routinely invoked by Enlightenment thinkers like William Paley and Voltaire.

229   **"As soon as the universe began":** Seth Lloyd, *Programming the Universe: A Quantum Computer Scientist Takes on the Cosmos* (New York: Knopf, 2006), 3.

229   **"All living creatures":** Charles Seife, *Decoding the Universe* (New York: Viking, 2006), 262–64.

Epilogue: Flame and Filament

231   **"as revolutionary in the development":** Schivelbusch, *Disenchanted Night*, 6.

232   **"the soul of the house":** Ibid., 28.

232   **"We have noticed":** Quoted in ibid., 178.

Acknowledgments

In addition to the writers of the works cited in the Notes, the author thanks Mike Sullivan and Jennifer Lozier at VeriCenter; Marc Benioff and Parker Harris at Salesforce.com; Bryan Doerr at Savvis; Adam Selipsky at Amazon Web Services; Vladimir Miloushev at 3Tera; George O'Conor at Oco; Greg Gianforte at Right-Now; Jonathan Schwartz and Greg Papadopoulos at Sun Microsystems; Uwe Wagner at T-Systems; Lew Moorman at Rackspace; Patrick Grady at Rearden Commerce; David Scott at 3PAR; Ron Rose at Priceline; David Greschler at Softricity; Chris Bergonzi and Alden Hayashi at the *MIT Sloan Management Review*; the Gleason Public Library and the Middlesex Valley Library Consortium; his agent, Rafe Sagalyn; his editor, Brendan Curry, and the staff of W. W. Norton; and, most of all, his family.

Index

Page numbers after 235 refer to notes.